MOLECULAR REALITY

A Perspective on the Scientific Work of Jean Perrin

History of Science Library

Editor: Michael A. Hoskin
Lecturer in the History of Science, Cambridge University

THE ENGLISH PARACELSIANS
Allen G. Debus
Professor of the History of Science, University of Chicago

WILLIAM HERSCHEL AND THE CONSTRUCTION OF THE HEAVENS
M. A. Hoskin
Lecturer in the History of Science, Cambridge University

A HISTORY OF THE THEORIES OF RAIN
W. E. Knowles Middleton

THE ORIGINS OF CHEMISTRY
R. P. Multhauf
Director of the Museum of History and Technology at the Smithsonian Institution

THEORIES OF LIGHT FROM DESCARTES TO NEWTON
A. I. Sabra
Reader in the History of the Classical Tradition, The Warburg Institute

MEDICINE IN MEDIEVAL ENGLAND
C. H. Talbot
Research Fellow at the Wellcome Institute of the History of Medicine, London

THE EARTH IN DECAY
A HISTORY OF BRITISH GEOMORPHOLOGY, 1578–1878
G. L. Davies
Lecturer in the Department of Geography, Trinity College, University of Dublin

THE CONFLICT BETWEEN ATOMISM AND CONSERVATION THEORY 1644–1860
W. L. Scott
Professorial Lecturer, School of Business Administration and Department of Mathematics and Science, American University

THE ROAD TO MEDICAL ENLIGHTENMENT 1650–1695
L. S. King
Professorial Lecturer in the History of Medicine, University of Chicago

THE SCIENTIFIC ORIGINS OF NATIONAL SOCIALISM:
SOCIAL DARWINISM IN ERNST HAECKEL AND THE GERMAN MONIST LEAGUE
Daniel Gasman
Assistant Professor of History, Douglas Jay College, City University of New York

WILLIAM HARVEY
AND THE CIRCULATION OF THE BLOOD
Gweneth Whitteridge
Lecturer in the History of Medicine, University of Edinburgh

FORCE IN NEWTON'S PHYSICS
THE SCIENCE OF DYNAMICS IN THE SEVENTEENTH CENTURY
Richard S. Westfall
Professor of the History of Science, Indiana University

THE VORTEX THEORY OF PLANETARY MOTIONS
E. J. Aiton
Lecturer, Department of Mathematics, Didsbury College of Education

MOLECULAR REALITY

A Perspective on the Scientific Work of Jean Perrin

MARY JO NYE

Visiting Assistant Professor of the History of Science, Oklahoma University

MACDONALD : LONDON
AND
AMERICAN ELSEVIER INC: NEW YORK

To Bob

First published 1972

Sole distributors for the United States and Dependencies
American Elsevier Publishing Company, Inc.
52 Vanderbilt Avenue
New York, N.Y. 10017

Sole distributors for the British Isles and Commonwealth
Macdonald & Co. (Publishers) Ltd
49–50 Poland Street
London W1A 2 LG

All remaining areas
Elsevier Publishing Company
P.O. Box 211
Jan van Galenstraat 335
Amsterdam
The Netherlands

Library of Congress Catalog Card Number 70 171234
Standard Book Numbers
Macdonald ISBN 0356 03823 8
American Elsevier ISBN 0444 19596 3

Printed and Bound in England by
Hazell Watson and Viney Limited,
Aylesbury, Bucks

Contents

List of Plates

Appearing between pages 52–3 and 68–9

Acknowledgments

I would like to express my gratitude, first of all, to Professor Robert Siegfried for introducing me to the History of Science and for imparting to me an enthusiasm for its pursuit. I would also like to thank the National Science Foundation and the University of Wisconsin for supporting me while the research and writing of this analysis were accomplished.

To the personnel of the Bibliothèque Nationale, as well as to those of Memorial Library at the University of Wisconsin, I would like to pay tribute for their friendly advice and assistance regarding research materials. I should also like to thank Claude Guédon for deciphering a particularly difficult letter for me and the Faculty Research office of the University of Oklahoma for the use of their xerox facilities.

The debt I owe Professor Francis Perrin is very great. Not only did he loan me unpublished materials for leisurely perusal, but he also gave me a number of published books and articles, and photocopied for me a number of letters in which I was especially interested. I appreciate very much the time he spent in conversation with me, and I am most grateful to his wife for providing a relaxed and enjoyable atmosphere for an evening's discussion.

I am particularly grateful to Professor Erwin Hiebert for his most careful reading of this text, for his critical comments and his most helpful suggestions for improving certain aspects of it. He also obtained for me several letters from the Einstein correspondence at the Princeton Institute for Advanced Study which I found useful. The direction of his own intellectual interests in the History of Science has greatly influenced my own, and I am most indebted to him for his assistance in a foundation of my knowledge of the discipline.

Finally, I want to acknowledge an especially deep debt of gratitude to my husband Bob. In his own pursuit of a knowledge and understanding of European history, he has profoundly influenced my own intellectual growth, and he has been both patient and helpful when in the past few years the household imperative has been replaced by that of the library. It may be that an individual's awareness and comprehension are in the end an individual thing, but as with most experiences, they are more joyous and more meaningful for having been daily shared.

Introduction

In attempting to understand the chemistry and physics of the nineteenth and twentieth centuries, it is readily apparent that a watershed of fundamental significance marks the years leading into our present epoch. Many of the discoveries and theories with which every novice and amateur reader of popular science is familiar—in stereotype, if not in detail—were elucidated during this era. These include the phenomena of radioactivity, X-rays, electrons and quantum theory, and the scientific revolution they helped create has been treated from numerous different approaches elsewhere.

Yet something is missing in many of these histories, or perhaps too much is assumed. What has struck me in a perusal of the literature on these topics is the tendency to assume what so many of the physical scientists of this pivotal period did not for one minute assume—the discontinuity of the matter which underlies visible reality. In looking back upon the discoveries and theories of particles, one perhaps fails to realise that the focus was not simply upon the *nature* of the molecules, ions and atoms, but upon the very fact of their existence. The alternative to a wave conception of cathode rays, for example, could not be couched in simple 'either-or' terminology, for the 'or' was not recognised by a sizeable segment of the scientific group as even a possible alternative.

In studying cathode rays in particular, I found that one of the more important experiments prior to the work of J. J. Thomson was ingeniously designed by the French physical chemist Jean Perrin. Upon further inquiry into his work, I discovered that nearly all of his subsequent experimental endeavour was aimed at the resolution of the indecision over the atomic-molecular hypothesis as an assumption or possible alternative in any given physical discussion.

In analysing the role of Jean Perrin in the eventual acceptance of this assumption among the outspoken majority of the scientific community, I have concentrated upon the period of experimental, theoretical, philosophical and popular science which climaxed with the Solvay conference of 1911 and with the publication of Perrin's book *Les Atomes* in 1913. Although Perrin himself continued to do research in various areas to further underline his basic conviction of the reality of molecules and to determine the parameters of the discontinuous microworld, the essential point was accepted by 1913, and those who still condemned the molecular viewpoint as pure speculation in 1913 continued to do so until their death.

This study is not, however, a straightforward analysis of theory and experiment. Necessarily, the problem is intertwined with both the philosophy and popularisation of science, as ideas on the particulate or ultimate structure of matter have always been of some interest both to philosophers and to the populace. In addition, since the watershed at the turn of the century marked a fundamental change in the essential working conditions of science—no one would doubt that a revolution in scientific conceptualisation was occurring—one is not surprised to discover a wholesale examination of the basic rules of procedure of science. Hence one finds that the entire question of molecular reality involved the problem of procedure and a redefinition of the aims of scientific endeavour. In particular, Perrin himself faced the formidable technical difficulties inherent in a credible development of probabilistic, statistical thermodynamics, as originally conceived by Boltzmann. In addition he had to work out his own solution to the criticisms raised by the logical positivists (Mach, Duhem, and others) of the assumptions at the base of physical science.

In the first chapter of this study, I have investigated the opinions of nineteenth-century chemists and physicists *vis-à-vis* the validity of the atomic-molecular hypothesis, and I have attempted to document in a general way the complexity of the issue in its many ramifications. I have pointed out, too, that the well-known experimental phenomenon of Brownian movement, which had been studied since the 1820s, could have suggested a resolution to many of the ambiguities and controversies, had the right perspective only been taken. One would especially have thought that physicists concerned with the kinetic theory of gases might have adopted a fruitful experimental approach, but this did not occur.

In the second chapter, I have discussed in broad terms the scientific and public career of Jean Perrin, dwelling especially upon the scientific and intellectual milieu in which he was educated and demonstrating how that environment of turmoil affected his professional goals. Much of the chapter discusses his early experiments and writings, nearly all of which were aimed at contributing to a resolution of the controversies about the procedures of science and at documenting the reality of the molecular world. The third chapter, then, is devoted to his work on Brownian movement, for which he is most renowned in scientific literature today. His discoveries and theories were not carried out in a vacuum, of course, and the theoretical work of Einstein and other mathematicians, as well as the experimental investigations of The Svedberg, Victor Henri and other experimentalists, is related to Perrin and to his viewpoint. One finds that for most of these scientists, a motivating factor in their work was the recognition that Brownian movement was a crucial test of molecular reality and of the general theories to which the atomic-molecular hypothesis led.

In conclusion, I have discussed the reception of Perrin's scientific experimentation and propagandisation on the subject of molecular reality, especially his work on Brownian movement, which climaxed in 1913 with the completion of a number of national and international conferences and the publication of *Les Atomes*. Though Perrin himself did not view his task as completed at that time, the question was no longer central to the basic working assumptions of scientists, and polemics on this question were no longer an impediment or impetus to the progress of general scientific conceptualisation. That Perrin's role was historically essential to this denouement cannot, in my opinion, be doubted.

Chapter One

The Nineteenth-Century Background

When the young student Jean Perrin matriculated at the École normale supérieure in 1891, the Daltonian atomic theory was more than eighty years old. Contrary to what one might expect, however, the education of the *lycée* student of Perrin's generation, even of one aspiring to a career in physics and chemistry, included little mention of the word 'atom'. In France, especially, the term 'equivalent' was still in vogue after extenuated battles between chemists and physicists for more than seventy-five years, and, in fact, it was only in the 1890's, as Perrin was beginning his post-*lycée* studies, that one no longer noted a tinge of sarcasm in references to the *atome* at the École normale.[1]

From the outset, the conflict over terminology in dealing with the substructure of matter had been a free-wheeling one, beginning with the first polemics among the chemists of England, Germany, France and Italy. With the development of thermodynamics and kinetic theory as new adjuncts of chemistry, the atomic quarrels expanded into the empire of the physicists and the new physical chemists, reaching a peak of fury at the very time when Perrin was deeply immersed in learning the discipline of physical chemistry. This situation exposed him not only to differences of opinion over the facts of science, but also to dissensions over the philosophic foundations of science, with profound implications for its future practical development.

Perrin's position in the controversies evolved from his education at the École normale amidst this environment of staunch polemics, and while he found the French physicists and chemists, on the whole, to be only lukewarm about atomic-molecular conceptions,[2] his own publications and experiments were later to play a large part in a conversion of their viewpoint. Much of his success lay in his understanding of both the philosophical and pragmatic issues under discussion, resulting in his demonstration that the discontinuity of matter was not a metaphysical hypothesis, as many of the anti-atomists claimed, but that it was a physical hypothesis capable of experimental verification. Though he attempted such verification through several different focuses of experimentation, the most convincing of his endeavours was the utilisation of the phenomenon of Brownian movement. Although widely known since 1828, the complete and proper explanation of Brownian movement had nevertheless eluded a large number of nineteenth-century biologists and physicists in the very midst of the polemics over the atomic-molecular hypothesis. Little hope had been offered even by the hypothesis' defenders for any visual verification, and this situation continued while the relevancy of Brownian movement was overlooked.

In attempting to understand the magnitude of Perrin's definitive establishment of the truth of molecular reality, it is imperative to examine the milieu in which his scientific intellect and experimental goals were formulated. Almost all of his publications, from the very outset of his career, were aimed at resolving the controversies of the nineteenth century, and he confidently tackled the ideological and experimental positions of some of the most prestigious figures in French science—including Marcelin Berthelot, Henri Le Chatelier, Pierre Duhem and (in a sense) Henri Poincaré—as well as such renowned *étrangères* as Wilhelm Ostwald and Ernst Mach. His utilisation of Brownian movement as the long-sought experimental confirmation of molecular reality also had roots in the nineteenth century, and his insight into the problem, piercing through the tangled skein of possible explanations posed by his predecessors, was an enormous achievement in itself.

The Initiation of Polemics: Dalton to Berthelot

The nineteenth-century polemics over the question of molecular reality began with the publication of Dalton's *System of Chemical Philosophy* in 1808. Dalton's fundamental premises of specifically different particles of matter for each element and of their combinations in multiple proportions were in part subsequently justified by Gay-Lussac's law of combining volumes (1808), Avogadro's hypothesis (1811), Dulong's and Petit's relation (1819), and Mitscherlich's theory of isomorphism (1819)[3] (although Dalton himself rejected many of these ideas).[4] At the same time, the atomistic conception was complicated by Prout's hypothesis (1815) of the hydrogen atom as a basic unit of matter contained in different proportions in each chemical element—a hypothesis which was refuted and then debated anew with the publication of diverse experimental figures for atomic weights.[5] In addition, Avogadro's law was not understood in its entirety, especially in its assertion that an element might consist in its gaseous form of two like atoms. Hence, William Whewell at mid-century questioned the validity of the atomic hypothesis in general. For example, in dealing with the combination of equal volumes of chlorine and hydrogen at a given temperature and pressure, he wrote that the resulting hydrochloric acid occupies the space of two volumes 'and therefore the proper number of particles cannot be supplied, and the uniform distribution of atoms in all gases maintained, without dividing into two each of the compound particles, constituted of an atom of chlorine and an atom of hydrogen'.[6]

An initial opposition to Dalton's atoms came from the pen of Humphry Davy, who insisted that the term 'atom' could only realistically have the meaning 'equivalent'—that the 'atom' was a unit of chemical reaction, rather than a material entity. Davy's opinion remained steadfast, even upon the occasion of presenting Dalton with the Royal Society's medal in 1826. Assessing Dalton's reputation in his address to the Society, Davy judged that this chemist's fame always would rest upon his discovery of that principle which fixes the proportions in which chemical bodies combine. By judicious use of tables of chemical equivalents, Davy cautioned, one might separate the practical part of Dalton's doctrine from the atomic or hypothetical part.[7]

The empirical laws which Davy distinguished from the hypothesis had been disengaged from the Daltonian atoms in the tables of Wollaston in 1814, where the equivalents were designated as the *supposed* atomic weights, using oxygen as a base.[8] Though Wollaston originally had hoped to find proofs for Dalton's hypothesis, he conceded to Berthollet and Davy that the choice of one atomic weight rather than another, which might be just as probable, was insupportably arbitrary.[9]

There were, of course, those who happily grasped the Daltonian hypothesis with very little criticism. Berzelius and Thomas Thomson, for example, thought the atomic theory so self-evident that its consequences were completely obvious after a first play of the mind, though they each differed somewhat in approach from Dalton and from each other. Whereas Thomson mixed the Daltonian atomic theory quite liberally with the Proutian hypothesis, Berzelius would have none of the latter.[10] In adopting a particulate view of matter and Dalton's empirical laws, Berzelius none the less steered clear of the onus of metaphysical speculation. He considered Dalton's conception of diverse geometrical configurations for different atoms a flight of fancy[11] and preferred to think of atoms more simply as all spherical in shape and all of the same size.[12] While Berzelius' alphabetical symbolism improved the technique of communication of Dalton's conception, and while he employed spherical particles in his electrochemical theory of combinations and reactions, Berzelius still regarded the basic hypothesis as unsatisfactory, for it simply could not explain enough, it did not 'embrace all'[13] as a fundamental scientific theory should do.

Within the field of organic chemistry arose some of the strongest arguments for the absolute reality of atoms. Liebig, for example, attached a great deal of practical value to the idea of atoms, becoming so immersed in the working out of these concepts that he later wrote, 'All our ideas are so interwoven with Dalton's theory, that we cannot carry ourselves back to the times in which that theory did not exist.'[14] In 1826 Gay-Lussac and Liebig published their discovery of the identity in composition of silver fulminate and silver cyanate (AgCNO), and this was followed by Wöhler's transformation of ammonium cyanate into isomeric urea (1828) and Berzelius' discovery of the identity of racemic acid and tartaric acid (1830).[15] This isomerism seemed inexplicable except on the grounds of real, structural molecules. Similarly, the doctrine of radicals, the principle of substitution, and the theory of types all lent some credence to a belief in the physical reality of atoms and molecules.[16] Such arguments were utilised in various debates of the period, as in the 1869 debate at the evening meetings of the London Chemical Society, where Alexander Williamson cited this doctrine of molecules and their valencies as a proof for atomism.[17]

Yet there was still a great deal of hesitation about the apparent submicroscopic world of discontinuous matter. Van't Hoff, after explaining isomerism in organic compounds through the hypothesis of a spatial tetrahedral arrangement of the carbon atom's valencies, wrote to Arrhenius that 'The representations themselves, atom, molecule, their dimensions, and perhaps their shapes, are after all something doubtful, as is the tetrahedron itself.'[18] Putting the matter more explicitly, Kekulé, who publicly opposed Liebig on this matter, wrote that

> The question whether atoms exist or not has but little significance from a chemical point of view; its discussion belongs rather to metaphysics.

> In chemistry, we have only to decide whether the assumption of atoms is an hypothesis adopted to the explanation of chemical phenomena. . . . From a philosophical point of view, I do not believe in the actual existence of atoms, taking the word in its literal significance of indivisible particles of matter—I rather expect that we shall some day find for what we now call atoms a mathematico-mechanical explanation which will render an account of atomic weight, of atomicity, and of numerous other properties of the so-called atoms.[17]

In France, the initial viewpoint towards the Daltonian atomic theory and its usefulness was hypostatised by Berthollet, whose country home was the focus of a great deal of important work in chemistry and physics at the turn of the century, and whose pupils included Gay-Lussac and Dulong.[20] Although Berthollet accepted implicitly the particulate structure of matter, as did most chemists around 1800,[21] he was more concerned with the forces between particles than with the particles themselves.[22] It was he who wrote the introduction to the French translation of the third edition of Thomson's *System of Chemistry*, criticising the Daltonian theory from two principal standpoints: first, a distrust of invisible entities inaccessible to experience, and secondly, an opposition to the currently firmest experimental justification for the theory, the laws of constant multiple proportions.[23] Since this French edition was widely read by scientists on the Continent, this caution on Berthollet's part must have significantly influenced the next generation of chemists.

Gay-Lussac, for example, despite the fact that his law of combining volumes was later used to bolster Daltonian and general atomic theory, was himself quite critical of Dalton's concept. As Maurice Crosland writes in a recent analysis,

> Dalton, in the Newtonian tradition, considered that matter consisted of atoms characterized by weight. For Gay-Lussac, in the Cartesian tradition, weight or mass was not so fundamental a property as extension or volume.[24]

So influential was this concept of volume rather than mass as a fundamental entity that Dalton referred in an unpublished paper of 1830 to the 'French disease of volumes',[25] a malaise which Berzelius and a majority of Continental chemists shared.

Of perhaps even greater influence upon the attitude in France towards the atomic-molecular theory was Jean-Baptiste Dumas, who, like Marcelin Berthelot in the next generation, wielded enormous authority and prestige due to experimental successes and an established academic chair. In fact, so overwhelming was the influence of these men and so pervasive was the acceptance of their interpretations of science that, in assessing the role of external circumstances in scientific discovery, the historian René Taton has written that 'if their power is extended too much, some individuals can so profit as to reduce their young adversaries practically to silence and thus check the progress of science. Such was the case of two *savants*, in other respects distinguished, Jean-Baptiste Dumas and Marcelin Berthelot, who at a given moment created a certain scientific dogmatism against which it was very difficult to struggle.'[26]

In his early researches in 1826, Dumas analysed vapour densities for atomic weight determinations, recognising the relation of the hypothesis of Avogadro to Dalton's atomic theory and assuming that in all elastic fluids under the same conditions, the molecules are at equal distances since they are present in equal numbers. Uneasy, however, with the conception of a large number of entirely different elementary atoms, Dumas was attracted by Prout's hypothesis and attempted to verify it, asserting an analogy between the observed differences in atomic weights of homologous organic compounds.[27] By the same token, after studies of the substitution of chlorine, bromine, and iodine for hydrogen in organic compounds, Dumas at first hesitated to follow the extension of his doctrine of substitution into that of types, as suggested by his pupil Laurent. Though he eventually was seen vying with Laurent over the question of priorities for the type theory,[28] he still remained doubtful of the reality of the atoms. Rather, speaking of the atomic theory itself, he said, 'If I were master of the situation, I would efface the word atom from Science, persuaded that it goes further than experience and that, in chemistry, we should never go further than experience.'[29]

The alternative of avoiding the use of the word *atom* had indeed become feasible by mid-century, following Berzelius' popularisation of the use of *atom* and *volume* as interchangeable expressions[30] and Faraday's discovery of the fixed proportionality between the quantity of electrochemical decomposition and the already well-known chemical equivalents. Leopold Gmelin had also adopted a system in which compounds were assigned formulas on the basis of the number of equivalents chemically present.[31] In France, Auguste Laurent and Charles Gerhardt then led the battle for the atom, with Adolphe Wurtz,

Charles Friedel, and Paul Schutzenberger numbered among their followers.[32] At loggerheads with them were Saint-Claire Deville and Berthelot, and in 1877 the conflict broke into open confrontation in the chambers of the Académie des Sciences.

This particular confrontation was, in fact, a French re-enactment of the international drama which had occurred seventeen years earlier at the 1860 Karlsruhe Congress, where a steering committee, including Wurtz (along with Kekulé, Cannizzaro, Kopp and others), drafted a series of questions about atoms, molecules, radicals and equivalents to be discussed by the distinguished gathering of about 140 chemists.[33] That Deville, for one, had not changed his mind since the congress was quite evident in his attack on Avogadro's law in 1877 in the chambers of the Académie. Wurtz, who had studied under Liebig and under Dumas, and whose work on ammonium compounds had confirmed Laurent's and Gerhardt's theory of types, had persisted conversely in defending the atomic theory in France since succeeding Dumas at the Faculté de Médecine,[34] and he did not hesitate to answer Deville.

This defence of atomism resulted in a discussion in May and June of 1877 between Wurtz and Berthelot concerning the question of whether chemical changes should be expressed by elements in equivalent proportions or whether the more modern system of atomic weights should be employed. Wurtz maintained that the previous use of equivalents, rather than of Berzelius' atomic notation, had slowed chemical progress considerably since 1840, particularly in understanding the reactions of gases. Berthelot, on the other hand, denied that progress had been affected by a notation that did not form the foundation of science. He also disagreed with Wurtz' espousal of diatomic gases, asserting that this was mere assumption and objecting to the combination of two like elementary atoms as a 'mystical conception'. Further, he protested that Avogadro's dictum was mere hypothesis, not law, while propositions involving equivalents fulfilled the definition of 'laws'. Berthelot's criterion for the admission of molecules and atoms to the laws of chemistry was expressed perhaps a little too succinctly in his rhetorical query demanding '[And] who has ever seen a gas molecule or an atom?'[35]

In his reply to Berthelot, Wurtz agreed that his notion of atoms and molecules was indeed an hypothesis, and in the book *La Théorie Atomique*, published two years later, Wurtz reiterated the point, though with important qualifications.

> There are two things here that it is necessary not to confuse, facts and hypothesis. We will retain the hypothesis as long as it permits faithfully interpreting the facts; grouping them, relating them to each other and predicting new things, as long as, in a word, it will show itself fecund. Hypothesis thus developed is raised to the rank of theory. We will attempt to demonstrate that it is thus for Dalton's conception, in exposing its origins, its progress, its consequences.[36]

Hence, the role of hypothesis and imagination in science and their conversion into laws or theories were sharply at issue in these discussions, and Wurtz' stand is fairly unambiguous. Berthelot, however, still condemned a few years later what he deemed 'the continual effort of the human *esprit* to go beyond any demonstrated explanation of phenomena as soon as such an apparent explanation has been attained, and to hurl itself further towards new fancies'.[37]

This is not to say that Berthelot wished to exclude entirely imagination and hypothesis from scientific reasoning, but his scepticism about their role and especially about their usefulness in laws and theories is obvious. In assessing Berthelot's opposition to atomism, it is most often suggested that his concern for immediate and tangible experimental proof and his almost instinctive mistrust for hypothetical constructions were attached at base to the *esprit* of Comtian positivism.[38] Certainly the idea that inspired all of Berthelot's research was (the Comtian) belief in the unity of the universe, of its phenomena and of its laws.[39] When, after Claude Bernard's death, Berthelot published Bernard's notes on enzymes—which ran counter to Pasteur's view that fermentation can occur only in the presence of life—Pasteur immediately accused Berthelot of complicity in a positivist conspiracy (with Renan, Taine, Littré, and Bernard) though Berthelot explicitly denied adherence to the school of Auguste Comte.[40]

Berthelot's opposition to the atomic theory, however, was based on more than strictly positivistic reasoning, and in fairness to him, and with awareness of the circumstances under which the physical reality of atoms and molecules was finally to be acceptable in the early twentieth century, it is instructive to consider a passage in his *Origines de l'Alchimie*. There, following a discussion of the atomism of Leucippus, Democritus and Epicurus, and a statement of nineteenth-century atomic theory (that which was based on purely chemical considerations), Berthelot wrote:

> Physical forces, as well as chemical forces, will allow testing in the atom only movements of the whole, without the possibility of internal vibrations; these can exist only in a system formed of several parts. It results then that there cannot be in the interior of an indivisible atom any reserve of hidden energy. Such are the rigorous consequences of the atomic theory.[41]

His objections to an ultimately indivisible particle—one without the potentiality for internal motion—suggest that Berthelot might, indeed, have accepted a real molecule or atom such as was finally posed by J. J. Thomson, Ernest Rutherford, and Jean Perrin. Although his principal objections to the atomic-molecular conception were methodological and criticised the systematic utilisation of an entity inaccessible to direct experimentation, one suspects that he might have been more favourably disposed to an invisible atom that was not yet an ultimate particle but was itself composed of even more minute sub-particles in motion. With such a structured atom, one could better understand the energy relations of chemical reactions; the indivisibility of Dalton's atoms severely strained Berthelot's credulity.

The Phenomenon the Chemists failed to notice

The twentieth-century molecule and atom are the results, in large part, of imaginatively conceived physical experiments. This stands in contrast to the nineteenth-century discussions of atomic-molecular reality, where any real experimental demonstration of the discontinuity of matter seemed totally ephemeral. Berthelot's contention that he had never seen a molecule could scarcely be disputed, and few of his opponents—certainly not Wurtz—dreamed of a visual verification. Oddly enough, however, the possibility for an experimental demonstration of the particulate nature of matter did exist although its full potentiality was not to be realised until the work of Perrin in the next century. By 1830, Dumas, as well as the English chemists Wollaston and Prout, had some of the necessary information at their disposal, had they grasped the appropriate correlation. Strangely enough, this information was not related to researches in chemistry, but to the investigations by biologists of the problem of reproduction in plants.

Both Wollaston and Prout were friends of Robert Brown, whose name was subsequently associated with the discovery of 'Brownian movement': the motion of exceedingly small microscopic particles (on the order of 10^{-3} mm diameter) in fluids. Because of their friendship, with Brown, the two chemists Wollaston and Prout were, in fact, among the first scientists to know the details of Brown's studies and the vein of his surmises.[42] Simultaneously, in France, Dumas was himself enthusiastically interested in biology and was an editor of the *Annales des Sciences Naturelles*. His co-editor was Adolphe Brongniart, who had been his bosom friend since their youth and a brother-in-law since 1826.[43] Brongniart had plunged into studies of plant reproduction in these years and in France, at least, his name became indissolubly associated with that of Robert Brown.

The observation of Brownian movement had certainly occurred before the nineteenth century, but was never distinguished from the movements of Leeuwenhoek's 'animalcules' until the technical advance of the achromatic lens was made.[44] When Brongniart began his investigations of the granules contained in pollen, he proceeded from the premise, also held by others, that these granules were actually spermatic animalcules.[45] He observed their movement in water—'a sort of rapid turning and translation like that of Monads and other infusoria animalcules'[46]—and mused that they changed relative positions 'as if by the effect of a sort of attraction and repulsion'.[47] Larger granules mixed in with them remained immobile. Brongniart also noticed that the intensity of particle movement increased at higher temperatures and he suggested that a rise in temperature, due to a plant's absorption of oxygen, might augment the vitality of its pollen, furnishing an indispensable condition for fertilisation.

In another paper, published in 1828, Brongniart replied to the criticism of Raspail, who maintained that the granular movements were not at all vital, but were effects of surface capillarity, liquid evaporation, and (or) the agitation of the air, the observer or the city upon the water of the preparation.[48] Enclosing a drop of pollen granules in a glass capsule made with small concave lenses and covered with a very thin sheet of mica, Brongniart observed, on the contrary, that the motion continued despite the impossibility of any evaporation or agitation of the surface of the water. In addition, he carried out the experiment by light from clouds, in order to avoid heating effects which might occur from direct sunlight. He further pointed out that placing the granules in an alcoholic preparation rather than in water should increase the particles' velocities if the phenomenon were due to evaporation; on the contrary, this experiment led to a complete cessation of movement.[49]

Robert Brown's first paper on Brownian movement appeared in the *Philosophical Magazine* in 1828 and was immediately translated for the *Annales*. His observations of the oblong and spherical pollen granules of *Clarckia pulchella* were most similar to Brongniart's, as were his conclusions that the movements arose neither from currents in the fluid nor from its evaporation, but belonged to the particles themselves.[50] Brown, however, extended observations to '. . . specimens of several plants, some of which had been dried and preserved in an herbarium for upwards of twenty years, and others not less than a century'[51] and discovered that their 'molecules' still exhibited the same vigour. Quite impressed by the tenacity of the phenomenon, Brown recorded that he 'was disposed to believe that the minute spherical particles . . . were in reality the supposed constituent or elementary molecules of organic bodies, first so considered by Buffon and Needham . . . and very recently by Dr. Milne Edwards'.[52] This revival of Buffon's almost metaphysical units was quickly cut short by Brown's discovery that the motion was also characteristic to suspended particles of inorganic substances—of simple earths, metals, rocks, and even a fragment of the Sphinx.

Noting Brown's observations of inorganic granules, Brongniart warned that these movements seemed 'moins constans' than those of the pollen granules.[53] Observing, too, that the most vivacious of the inorganic movements occurred with the better conductors of electricity, he suggested that 'It is a lot more probable that these movements have repulsions or attractions of molecules for their cause, influences which are of the province of physics'.[54] Yet the possibility of electrical influences causing the motion of the spermatic pollen granules seemed questionable; their apparent changes in position could be due to some lens effect. 'The only thing on which I cannot maintain any doubt', Brongniart continued, 'and on which I have the happiness of seeing my opinion entirely confirmed, by that of the commissaries of the Académie and of Mr Brown, is the complete independence of the movement from all exterior causes influencing the ambiant liquid': the movement was inherent to the granules themselves.[55]

With these viewpoints, then, the stage was set for numerous discussions in the next few years over the cause of the Brownian motion, with the premise of 'animalcule' movement gradually discarded in favour of various mechanistic interpretations. Brown himself was dismayed with Brongniart's hypothesis of electrical attractions, and devised an experiment in which he dissolved a drop of a particle suspension in a larger drop of almond oil, producing minute drops of water of various sizes, some containing only one particle. In each drop, the motion of the particle or particles continued undiminished in activity, in a situation where evaporation or mutual attraction was very much reduced or nil. After thus refuting the hypothesis of electrical attractions, Brown also experimentally discounted any explanation by capillarity.[56]

In the meantime, Raspail persisted in his opinion that the motions might easily be explained by the concurrence of all kinds of mechanical effects, that it was 'unnecessary to adduce individual facts after the general law has been ascertained'.[57] Similarly, Muncke (at Heidelberg) studied suspensions of gamboge—a yellow vegetable latex—and attributed the movement to unequal temperatures of strongly lighted water, to evaporation, to heat currents, and to other mechanical causes.[58] David Brewster suggested a new explanation (which never really gained favour) that the movement was due in part to a polarity in the granules: Brewster's spreading a large square of glass with a milky suspension did not, however, produce the anticipated interaction with polarised light. Yet, wrote Brewster, '... it is the last supposition in philosophy that they [the movements] are owing to animal life.... Why should not the molecules of the hardest solids have their orbits, their centres of attraction, and the same varied movements which are observed in planetary and nebulous matter?'[59]

A more intriguing explanation appeared in 1831, though it was not at all developed by the author and was totally ignored by the biologists, chemists, and physicists. Wrote M. Marx, a professor at Braunschweig, following observations of a gamboge suspension and its apparent independence from all exterior influences, 'It appears that the phenomenon belongs to a particular mode of action of the water on the resin particles'.[60] Similarly, in his treatise on microscopy, Félix Dujardin dismissed light, electricity, and magnetism as possible culprits for the activity, and postulated that it was the result of 'various impulsions that each particle receives from part of the radiating caloric emitted by all the neighbouring bodies'.[61] Clearly the whole question was bordering on either physics or metaphysics, and the subject had completely escaped the boundaries of biology. In fact, the methods for the solution to this problem were to be found neither in chemistry nor biology, but in the discipline of physical chemistry. Although this new science was initially to ignore the existence of the strange phenomenon, it indeed eventually recognised Brownian movement as one of the keys to the fundamental riddles of physical chemistry itself.

The Positivist Critique of the Atomic-Mechanical Picture

It was, of course, Jean Perrin who most fully explored the potentiality of experimental Brownian movement to unravel some of the traditionally perplexing problems of physical chemistry. As a student in the French university establishment at the turn of the century, Perrin was to study carefully both the research achievements and the epistemological arguments concerning atomism and other vital questions within physics and chemistry. A majority of these accomplishments and polemics of the nineteenth century occurred outside his own country, and it was, in fact, in his own generation that a focus of research and discovery returned to France. Influenced in both his reading and in the lecture room by the intellectual traditions of England, Germany and Austria, Perrin's own research interests, epistemological views, and personal goals were—as physical chemistry itself—more than exclusively French.

By the end of the nineteenth century, Jean Perrin was studying a discipline of physical chemistry which had reached a stage of development bridging what had once seemed an enormous gulf between the two separate sciences of physics and chemistry. By mid-nineteenth century both the first and second laws of thermodynamics had been successfully stated in various forms by an assemblage of scientists—the former by Colding, Mayer, Joule and Helmholtz, and the latter by Sadi Carnot, William Thomson, Clausius and Rankine. In the next years the application of the newly elaborated laws was made to numerous physical and chemical problems amidst the simultaneous description (Maxwell) and experimental verification (Hertz) of the electromagnetic character of light. This was the period of the development of the kinetic theory of gases by Maxwell and Boltzmann, with its concomitant statistical interpretations of the second law of thermodynamics, modifying Clausius' original definition of entropy. Equally sharing the limelight in these years of brilliant theorisation were the mathematical achievements of the American chemist, Josiah Willard Gibbs, in his application of thermodynamic principles to chemical equilibria.

Such remarkable fecundity in posing new scientific theories was certainly not accompanied by complacency among philosophers and epistemologists of science. In particular—in regard to the problem of scientific hypotheses, the atom and molecule as specific hypotheses, and the description of their relationships in mechanical terminology—the period of the development of physical chemistry witnessed a thoroughgoing inquest into the validity of a large number of traditional assumptions. Controversies over the kinetic theory ignited into a veritable bonfire, with those individuals who had been previously committed to the idea of the discontinuity of matter from the standpoint of pure chemistry now standing as firm adherents to the kinetic theories of Maxwell and Boltzmann. Cannizzaro, for example, wrote of the kinetic theory that 'In this manner alone can all that relates to the molecular and atomic constitution of bodies be deduced from a single principle, which is, indeed that most natural passage from physics to chemistry. Is it necessary for me here to declare that I do not admit that difference between the physical molecule and the chemical molecule which some persons have wished to introduce?'[62] Wurtz, too, insisted that 'As heat itself is a mode of movement, it results that the thermochemical facts are perfectly adapted to the atomic hypothesis.'[63]

Yet the classical interpretations of the first and second laws of thermodynamics suggested an entirely different conclusion to a large number of scientists, and the pursuit of physical problems, such as the observation of the critical point and liquifaction of gases, along with studies of chemical equilibrium and the application of Gibbs' phase rule, Le Chatelier's principle, and Duhem's thermodynamic potential, only confirmed a tendency to view matter as continuous in nature and to question critically the particulate, mechanical models of the past. In fact, the situation was such that in his essays on chemistry, published in 1894, T. E. Thorpe wrote that

> Corpuscular theories are now altogether banished from certain domains of physics. Indeed, the most weighty attacks yet made on the atomic theory have been delivered by those who are mainly occupied with the problems and abstract conceptions of energy. It is significant that Faraday, who began his scientific career as a chemist, seems to have gradually loosened his hold on the atomic theory as he became more and more absorbed in the contemplation of purely physical phenomena.[64]

Though Faraday himself did not have the mathematical competence or ability, his work indeed made possible that trend which resulted in the substitution for discontinuous models of sets of differential equations.

The best example of this trend was James Clerk Maxwell's improvisation upon the experimental researches of Faraday into the electromagnetic theory of light. This theory was proposed in three treatises, two containing elaborate mechanical analogies for an electromagnetic field, the most controversial one utilising a layer of particles, acting as idle wheels (ball bearings) between rotating vortices. After developing equations from these analogies, which he insisted were meant to suggest nothing of reality, Maxwell freed the equations from their model in his third treatise. Hertz' experimental verification suggested the profitability of such a scientific method, and the British school in general began formulating differential equations to demonstrate that the most varied phenomena—such as transmission of heat, propagation of electricity in conductors, and hydrodynamical problems—could be explained in the very same way. Soon the older theories themselves, such as the wave theory of light, theories of gases, and the geometrical schemes of the chemists, were considered simple mechanical analogies, and the philosophy of Maxwell was generalised into the dictum that knowledge is nothing other than the discovery of analogies.[65]

The development of this viewpoint strode hand in hand with the epistemological views of Heinrich Hertz. Hertz, who had been Kirchhoff's pupil at Berlin, attempted a reconstruction of mechanics upon the basis of mass, time, and space, rejecting the concept of force as a foundation for mechanics. He charged that the mechanists' 'force' lacked clarity due to scientists' introduction of unessential and hypothetical characteristics to its essential content. He himself was concerned not so much with the verification of individual (and hypothetical) entities, symbols or theorems, but with the verification of a system as a whole. His famous response to the question 'what is Maxwell's Theory' displays this interest: 'I know of no shorter or more definite answer than the following: Maxwell's Theory is Maxwell's system of equations.'[66]

Such a viewpoint logically suggests two possible alternatives in method for physicists and chemists in the nineteenth century. The use of mechanical analogies—employing invisible entities such as vortices, hidden masses or atoms—could be used to formulate differential equations and their validity experimentally verified according to their ability to predict physical results. A variety of different models might easily result in identical physical measurements, in which circumstance no judgment could really be made upon the relationship of any one model to physical reality. Such a method was followed by the English school of physicists almost in entirety, and it was no happenstance that Boltzmann's kinetic theory of gases was most avidly accepted by those scientists who were most enthusiastic about the application of mechanical analogies.

On the other hand, a second reaction to the methodology implied in Maxwell's theory (and later expostulated by Hertz) was to insist upon the unmitigated destruction of any mechanical analogy once its differential equations had been verified; one must never again be tempted to refer to it. Furthermore, the goal was set for the method of mechanical analogies to be discarded in the future. Although, historically speaking, most of chemistry and physics had been aided by the utilisation of mechanical models, such a course had not been logically inevitable, and if one could avoid mechanical models in the future, the painful contemporary questions about scientific knowledge of physical reality would simply never arise. In its crudest form, such a scientific methodology was espoused by Wilhelm Ostwald, and in its more sophisticated versions by Ernst Mach and Pierre Duhem, all of whom were primarily interested in thermodynamics. Each of these adherents was also a confirmed opponent of the atomic theory, with Ostwald and Mach publishing vicious polemics against Boltzmann on the subject.

This is not to say that an opposition to the atomic hypothesis or to molecular reality was logically necessary to someone who opposed the validity of mechanical models (the engineer's universe) and who opposed the use of hypothesis in a deductive approach to scientific theory. Historically, however, these problems were often interlinked, and the atomic-molecular hypothesis seemed the best example of a metaphysical proposition at the base of a large number of scientific theories which were in some cases—as with the kinetic theory of gases—unsatisfactory because they were apparently only partially 'true'. That this fundamental invisible entity also was capable of threatening the absolute character of thermodynamics could only add to the venom of those individuals who, beginning from an experiential-positivist attitude towards knowledge, had derived the differential equations of the thermodynamic edifice. It is in this context, then, that the concept of molecular reality became central, and a rejection of this hypothesis was intricately entwined with a rejection of kinetic theory in particular, and with a failure to understand the nature of matter in general. It was, of course, in this complex milieu of experiment, theory and philosophy that Jean Perrin was to operate. Indeed, as will be seen, most of his own experimental endeavours and epistemological views were in direct reaction to the general controversy and were strongly in sympathy with Boltzmann.

Ostwald, who was one of Boltzmann's most vociferous opponents, had been profoundly struck, it appears, by the fact that most of the terms in Gibbs' thermodynamical equations represented energy in different forms. Gradually he became more and more convinced that molecules, atoms, and ions were only mathematical and *a priori* fictions and that the real underlying component of the universe was energy in its various arrays.[67] If one designates under the name of 'substance' that which persists through changes of circumstances, he wrote, then the experience of work [energy] is more logically 'substance' than those concepts of mass and weight historically designated as matter.[68]

Ostwald's doctrine of 'energetics' was a challenge to the traditional methods of Newtonian mechanics and to the use of any *a priori* hypotheses, though his objections against the atomic theory may have been directed more against the mechanical picture in general, and the engineer's universe which it constructed, than against the existence of atoms themselves.[69] While admitting that the atomic hypothesis had been fecund in the early stages of chemistry, he asserted that it was now only posing pseudo-problems—statements which were only questions in the sense of hypothesis and for which it was impossible to experimentally demonstrate any real content. Taking an essentially positivistic viewpoint, Ostwald wrote that knowledge of nature must proceed directly from experience.[70] While he admitted that the atomic hypothesis was indeed historically responsible for the deduction of the early stoichiometrical laws of chemistry, Ostwald asserted in 1904 that even these may be arrived at through thermodynamic, experimental considerations of phase equilibrium. 'Chemical dynamics has . . . made the atomic hypothesis unnecessary for this purpose and has put the theory of stoichiometrical laws on more secure ground than that furnished by a mere hypothesis.'[71]

Ostwald's purge of mechanical hypotheses from physical theories was greeted sympathetically by a large number of figures in chemistry and physics, some of whom consented only to his methodological approach, while others, such as Georg Helm, indiscriminately adopted the whole of his energetic doctrine. Van't Hoff confided to Arrhenius that 'You write about the anti-kinetic utterances of Ostwald. I, too, must say that with a fairly large expenditure of mathematical development the kinetic theory barely gives the current 4% interest on capital, and I think that even this theory should be measured by its fruits.'[72] Van't Hoff leaned towards energetics as a form of exposition because it did exclude hypothesis.[73]

Typical, perhaps, of the leading physicists' responses to Ostwald's methodological critique and his doctrine of energetics was that of Max Planck. In Planck's papers on entropy during the period 1887–1891, for example, he proceeded without regard to questions concerning the ultimate structure of matter, utilising a classical thermodynamics that was currently providing quite successful solutions to a number of important scientific questions. While voicing dismay at the doctrine of energetics *per se*, Planck, like van't Hoff, opined that atomistics and kinetics *had* been leading a rather sterile existence.[74] By 1897, Planck was still avoiding references, in his writings and lectures, to kinetic-molecular theory,[75] and in the first edition of his textbook of thermodynamics he wrote of the kinetic theory that:

> Obstacles, at present unsurmountable, however, seem to stand in the way of its further progress. These are due not only to the highly complicated mathematical treatment of the accepted hypotheses, but principally to essential difficulties, not to be discussed here, in the mechanical interpretation of the fundamental principles of Thermodynamics.[76]

With this foreword, Planck then explained that his textbook utilised the 'inductive method' because it 'corresponds best to the present state of the science'. He cautioned, however, that affairs might change quite a bit in the course of the next few years and that a mechanical or perhaps electromagnetic conception might even now be in the ascendancy.[77]

Similarly, the criticisms expressed by Ernst Mach of kinetic theory and the mechanical picture in general, as well as of atomism in particular, were compatible with those of Ostwald, although Mach remained aloof from the doctrine of energetics. His objection to the abstract notion of matter at the base of traditional mechanics sprang from a rejection of the distinction between primary and secondary qualities, rather than from an affirmation of the primacy of energy.[78] Concepts such as mass and force, he maintained, were merely 'auxiliary concepts' by which statements about sense observations could be expressed in convenient fashion. They do not indicate entities of a real world hidden behind appearance.[79] Physics was, by definition, experience arranged in economical order, profitably employing the fact that the relations between experimentally observed quantities are similar to certain relations obtained between familiar mathematical functions.[80]

Regarding the role of hypothesis, Mach warned that 'Once an hypothesis has facilitated, as best it can, our views of new facts, by the substitution of more familiar ideas, its powers are exhausted. We err when we expect more enlightenment from an hypothesis than from the facts themselves.'[81] Once the scaffold has served its ends, it is to be quickly removed; hypotheses are only a preparative step towards a descriptive physics in the spirit of Maxwell's electromagnetic theory.[82] Strongly objecting to what he considered the experimentally unverifiable nature of atoms and molecules, Mach wrote that the postulates of the atom and molecule were based upon visual abstraction and yet they were defined from the outset as invisible entities. Further, the idea of an indivisible body—the atom—contradicted all actual physical experience and was intellectually a much less economical concept than that of the continuity of matter.* The principle of continuity, he wrote, achieves a simple result with the smallest amount of effort.[83]

Though one could not say that Ostwald, Helm, Planck, and Mach formed a coherent party or camp, it did happen that by the 1890's they each disagreed on one point or another with Boltzmann. As the public defender of the utility of the molecular-atomic hypothesis, the statistical interpretation of kinetic theory, and the probabilistic definition of entropy, Boltzmann gave to thermodynamics a mechanical form which was simply too all-encompassing for many of his contemporaries, and a distinct pole of opinion formed against him.

Boltzmann himself deplored his opponents' methodology for its lack of sensitivity to historical development.[84] The phenomenologists, he

* In comprehending what is meant by the continuity of matter, a general Cartesian picture is most often cited in the nineteenth-century literature. For example, in *La Théorie Atomique* Adolphe Wurtz describes the continuous universe as one in a state of permanent vibratory movement, with waves propagated in diverse directions and crossing each other. From the intersection of the waves result nodal surfaces and points, and consequently, limited portions of matter. The diversity of matter results from the diversity of the systems of waves, and the portions thus delimited—these *tranches vibrantes*—represent the particles of matter which enter into conflict in chemical reactions.

said, had purged themselves of atomism by depending upon differential equations and what they deemed the implicit continuity of matter characteristic of these equations. Yet 'the conceptions of an integral and differential calculus released from all atomistic representation . . . are purely metaphysical, if we understand by that—according to a famous definition of Mach—things which we have forgotten how we arrived at'.[85] The differential equations and calculus were attained, Boltzmann reminded his opponents, by taking the limits of discontinuous points: 'Atomistics appears inseparable from the very concept of continuity'.[86]

Although Boltzmann never asserted unequivocally the physical reality of atoms and molecules, he was convinced of the necessity of hypothesis in general, and of the usefulness of this one especially. For him, hypothesis was image or model (*Bild*) in a sense both simultaneously abstract and precise, yet always at odds with strict phenomenology in the sense of the cataloguing of facts of experience.[87] While the fashion seemed to be turning to an experimental, descriptive systemisation of physics, Boltzmann confessed in a speech at the University of Graz in 1890,

> According to my feeling, the task of theory lies in the construction of an image of the exterior world, which exists only in our mind and should serve to guide us in all our thoughts and all our experiments . . . . The more abstract theoretical research is, the more powerful it is.[88]

The final arbiter, however, to such deduced construction was experiment: 'If it is afterwards shown to be in accord with experiment, the arbitrariness of the foundations of our representation will be excused',[89] and it was only in this context that Boltzmann so vociferously defended the aged hypothesis of the discontinuity of matter. He deemed that it had not disappointed the stringent criterion of experiment.

In defending his own physical interpretations, Boltzmann not only dealt with his opponents' criticism of the kinetic theory of gases,[90] but also cited a number of positive results which had most recently accrued from the atomic-molecular theory. Van der Waals, for example, had established a formula for the behaviour of fluids by proceeding from purely molecular assumptions to entirely new results in the theory of corresponding states. Recent improvements in his formulation had produced precise representations of the behaviour of simple chemical substances, such as argon and helium, in such a way that atomistics had approached very closely to the phenomenologist ideal of expressing all states of bodies by one mathematical formula. Further, Boltzmann catalogued, atomistics had concretised and extended the dissociation theory of Gibbs, whose thermodynamic assumption, that in a dissociated gas all constituent parts behave in independent fashion such that their entropies, pressures, etc., can be determined, makes sense only when viewed from the standpoint of molecular mechanics. A further triumph for atomism consisted in Clausius' estimation of specific heats, as verified for mercury vapour by Kundt and Warburg and for rare gases by Rayleigh and Ramsay.[91]

Yet, in Boltzmann's catalogue of experimental supports for the discontinuous hypothesis and the mechanical and kinetic models, one notable experimental demonstration remained absent. Perhaps the phenomenon of Brownian motion seems to find its niche only by hindsight, especially considering the state of its comprehension as indicated in investigations prior to the middle of the nineteenth century. In the very issue, however, of the *Revue générale des sciences* where Ostwald's 1895 article 'La déroute de l'atomisme contemporain' appeared, there was also published an article by the French physicist Léon Gouy, entitled 'Le mouvement brownien et les mouvements moléculaires', and indeed the conclusions of Gouy's experimental paper might well have been startling if the right eye had then fallen upon them. As Gouy himself was to write a number of years later in correspondence with Jean Perrin, 'From the historical point of view, one wonders today how the great founders of kinetic theory, Clausius, Thomson, etc., have not been able to see that Brownian movement places under the eyes the realisation of all their hypotheses! *oculus habent*. . . .'[92]

What, then, is the Nature—or Relevance—of Brownian Movement?

The initial response to a physical interpretation of the phenomenon of Brownian movement was most often to regard it as irrelevant. Certainly, all opinions were not as cynical and extreme as that of the Reverend John Henry Conybeare, professor at Oxford, who wrote of Robert Brown's critical probe into the phenomenon,

> I don't believe a word on't . . . a thousand circumstances may generate motion . . . if, however, they should be confirmed, I know nothing inconsistent with the received philosophical notion as to the intimate corpuscular structure of bodies in them. Biot . . . states it to be possible that solid bodies may be compared of [sic] systems of moving molecules, representing in small what the planetary systems do in large. I would

> only add one supposition more; that these molecules are inhabited, and have philosophers among their population who, having observed the motion of some half-dozen molecules in their neighbourhood and ascertained their laws, believe they have developed the system of the universe.[93]

Yet those physical scientists who did attempt to correlate the frenzied particle motions with the molecular constitution of liquids encountered a great deal of opposition, and Maxwell himself supposed the movement only due to some freak agitation from the external surroundings; 'submitted to a more powerful microscope,' he opined, 'the bodies will demonstrate only more perfect repose'.[94]

One of the first physicists to treat the new problem was Christian Wiener, who published in 1863[95] a paper with the explicit purpose of explaining the atomic constitution of liquids, using the Brownian molecular movements (*Molekularbewegungen*). After eliminating the argument that the movements were those of infusoria (by heating quartz particles to a red-hot temperature), he also refuted a dependence upon electrical considerations (since particles very far apart behaved even as a large number of particles close together). He suggested that temperature differentials in the fluid were inadequate to explain the observations since temperature differentials would produce currents from the surface towards the interior, but the movement of the particles was not along these lines and its direction in a small volume was constantly changing. Similarly, evaporation proved an inadequate explanation since the movement proceeded identically in droplets with extremely different rates of evaporation.

With the most probable exterior causes eliminated, Wiener then attributed the movements to the make-up of the liquid itself ('... es bleibt uns daher Nichts übrig, als die Ursache in der Flüssigkeit an und für sich zu suchen, und sie *inneren*, dem *Flüssigkeitszustande eigenthümlichen Bewegungen* zuzuschreiben').[96] Unfortunately, however, for the progress in understanding Brownian movement, his conclusions were intertwined with a theory of aether-and-matter atoms, whereby the heat of a body consisted in the active condition of both kinds of atoms.[97]

Wiener's interpretation, linking Brownian movement to the internal molecular constitution of liquids, was met by a flurry of activity countermanding his pronouncement. In the next forty years, biologists and physicists prepared a variety of organic and inorganic particles—such as sulphur, mastic, cinnabar, pulverised coal, india ink, and gamboge—in organic and inorganic solvents, studying the influence of alkalis, acids, salts, and even narcotics upon the movement.[98] They subjected preparations of Brownian particles to sunlight and darkness, magnetism and electricity, red and blue light, and heat and cold in hopes of discovering the cause of the motion.[99] In 1886, for example, the *New York Microscopical Society Journal* recorded that one experimenter observed that the Brownian movement of globules in freshly drawn human milk varied, being most active at the birth of the offspring and diminishing in rapidity with an elapse of time.[100]

One of the most widely accepted theories of the movements was that of Sigmund Exner, who concluded that they were due to small currents in the liquid, emanating either from heat or light radiations. Exner remarked that one often viewed particles approaching each other without cohering, and such an occurrence could be explained only if the particles swam in different currents. These currents lent part of their own kinetic energy to the particles under their influence, accounting for the trembling appearance of the particles. An increase in particle activity accompanied a corresponding change in the intensity of heat or light, he further observed, and even the quality of light seemed to have some effect, with blue producing a greater particle displacement than red.[101] In England, J. B. Dancer concurred with Exner's judgment,[102] and Walter N. Hartley extended the idea of the influence of heat to argue that gamboge particles, as well as minute bubbles in mineral cavities,[103] owed their movement to an alteration in the surface tension of the surrounding liquid due, in turn, to an alteration in temperature of the medium. The change in surface tension then propelled any particle or bubble towards the source of heat.[104] By the turn of the century, there was a strong segment of opinion favouring some explanation through the role of heat. Wrote G. Quincke of Heidelberg, 'The Brownian molecular movements disappear, as Exner has shown, as soon as the liquid and its surroundings have the same temperature.'[105]

In competition with this theory of temperature differentials were a number of other explications, one, for example, basically centred on the argument of electrical influences. One of the strongest proponents of this outlook was William Stanley Jevons, who had observed that the granular movement was inhibited by the addition of most acids, alkalis, or salts, and who compared this inhibition to the prevention of the production of electricity in a hydroelectric machine. The addition of a salt, he wrote, prevents hydroelectrical actions, and it also prevents Brownian movement; the *experimentum crucis*, for Jevons, lay in the fact that ammonia and boracic acid were completely impotent in hindering the Brownian particle motion, even as they had no effect upon the operation of an electric machine. Jevons also drew an analogy between Brownian movement and electrical osmosis.[106] In addition to Jevons' hypotheses, a third explanation of the 'molecular movements', expressed by M. C. Maltezos, concluded that Brownian movement was a phenomenon of capillarity, dependent upon the inhomogeneity of the surface density of the individual suspended corpuscles, or perhaps upon the impurity of the liquid near the corpuscles.[107]

One must not assume, however, that the kinetic explanation was completely discarded during these years. In fact, the very length of the period in which experimentalists had observed the phenomenon was a kind of evidence for the kinetic interpretation; otherwise it was most difficult to account for the continued activity of Brownian particles in individual slide preparations which had, in some cases, been conserved and studied for almost twenty years under all types of external conditions.[108] Indeed, in 1877, the Brownian phenomenon was explicitly linked to the modern kinetic theory of gases and thermodynamics, as well as to the exciting new experiments of William Crookes on the particulate nature of matter in general. This was done in a paper by Joseph Delsaulx,[109] whose stated purpose was to demonstrate that the Brownian movements proceeded 'necessarily from the molecular heat motions, universally admitted, in gases and liquids by the best authorised promoters of the mechanical theory of heat.'[110]

Clausius, Delsaulx explained, had attributed a mean velocity to all molecules in a gas, so that their pressure at any point on the wall of a containing vessel would be the same. Yet this could only be true where the dimensions of the vessel were much greater than the mean free path of the molecules. For the particles and bubbles observed in Brownian motion, both the molecular and particular (bubble) dimensions are of the same order, and consequently the pressures at different points of a given particle or bubble are unequal; at different instants, the pressure even differs for the same point. Using Crookes' and Finkener's radiometer experiments relating the size of the radiometer discs and the mean free path of gas molecules, one can calculate that an observable effect should occur for the interaction of the dimensions of Brownian particles or bubbles and the even smaller mean free path of liquid molecules. 'If the molecules in the mass of liquid are only subject to single vibratory motions,' Delsaulx concluded, 'the displacing of the centre of gravity of the Brownian particles, consisting of thousands of molecules, would always remain imperceptible. [But] the influence of the motion of translation of the liquid molecules acting on the surface of the particles, permits us to explain easily all displacements hitherto observed.'[111]

Similarly, although in a much more sophisticated and thoroughgoing fashion, J. Thirion published a lengthy article in 1880, in which he utilised Brownian movement as a visible example of the mechanical theory of heat and the atomic-molecular hypothesis. Devoting most of his exposé to the cathode-ray experiments of Crookes, Thirion turned to Brownian movement to demonstrate in liquids what Crookes had demarcated in gases. Describing the completely irregular motion of liquid molecules and their mean or compensating pressure on the walls of their container, he sought a social analogy to explain the effect of molecules on large particles, in contrast to their effect on large walls:

> It is thus that statistics reveals more and more the inconstancy and the irregularity of much of social phenomena, when in lieu of applying it to a great nation altogether, one descends to a province, a town, a village. It should then be sufficient to isolate portions of smaller and smaller surfaces of evaporation or of pressure, in order to arrive finally at dimensions incapable of assuring compensations of irregularities, and then the effects of discontinuity and of the thermodynamic agitation of the liquid should be made sensible.[112]

All particles, Thirion explained, which are freely suspended in a liquid should oscillate without ceasing if they are sufficiently small, according to a strict interpretation of the mechanical theory of heat, and the smaller the dimensions, the larger the oscillations.[113] Had physicists only realised earlier the import of Brownian movement, he lamented—had they not blamed the phenomenon on errors introduced into microscopical research—progress in physics and the kinetic theory would have been much accelerated, confirming the mechanical theory of matter.[114]

Such a bright conjunction of the dancing Brownian particles and the controversial kinetic-mechanical theory did not go unchallenged. The strongest refutation proceeded from the German cytologist Karl von Nägeli, who assaulted the mechanical theorists on their own terms, calculating the probable velocities which would be produced by collisions of gas molecules (whose velocities and kinetic energies were already predicted from kinetic theory) and dust particles. He then extrapolated these results to the problem of minute particles in water, assuming the velocity of water molecules to be smaller than that of gaseous molecules and the density of the medium to be much greater. His conclusions were negative, predicting that a million water molecules must strike the starch particles (*Stärkekörnchen*) at the same moment and from the same direction in order to explain the single jerk of the Brownian granule. Certainly, he admitted, there are probably more than a billion molecular collisions against any one tiny particle in a given second, but the impulses come from all different directions. Nägeli's allegiance lay, rather, in the compromise viewpoint that currents, prompted by temperature differentials, accounted for the Brownian phenomenon, with an auxiliary role played by electrical attractions between solid particles surrounded by a mantle of either the suspensory liquid or of other molecules soluble in the principal suspensory liquid.[115]

Such a presentation was difficult to counter, and one of the few scientists who attempted to do so in the next few years was William Ramsay. Like several others before him, Ramsay first eliminated various external influences as causes of the phenomenon and noted its dependence only upon the size of the particles, their densities, and the nature of the liquid. He assumed that the motion was not due to currents in the liquid because it was so completely irregular in character. 'If molecules do not coalesce and move as a whole, then they would appear to have no possible power of giving motion to a mass so much larger than themselves,' he conceded to Nägeli, 'but that molecules have arrangement is probable, owing to the power which some liquids possess of rotating the plane of polarised light.'[116] Almost ten years later, Ramsay further elaborated, with a great deal of insight, but with not quite enough understanding of Boltzmann's statistical interpretations:

> Such a particle [of mass $2{\cdot}8 \times 10^{-12}$ gm, diameter $1{\cdot}4 \times 10^{-4}$ cm] has one hundred billion times the estimated mass of a water molecule; hence, if its motion be produced by bombardment from water molecules, these must exist in complex groups of considerable mass, and of some stability. It is very unlikely that pedetic motion is the result of electric charge on the particles, because they appear to be uninfluenced by each other's motion; there is no obvious source of such electric poten-

> tial, and moreover, we know from Messrs. Linder and Picton's experiments [on coagulation] . . . their pedetic motion is not interfered with. . . . The fact that pedesis is stopped by the addition of an electrolyte would appear to show that the water complices are disintegrated by the presence of ions. . . . The effect of pedetic motions in a liquid is to cause hydrostatic pressure. . . . It would appear not in the least unlikely that equal numbers of equally sized particles would exert equal pressure, as in the case of gases. . . .[117]

At the same time that Ramsay was publishing on Brownian movement, the French physicist Léon Gouy undertook a series of experiments. Gouy's work had heretofore dealt with optics (diffraction, reflection and refraction), thermodynamics and capillary phenomena, and he had closely followed Crookes' cathode-ray experiments. His first observations on Brownian movement were similar to those of everyone else, using gamboge and india ink in water, and observing the irregular, apparently independent trepidations with a magnification of 500 diameters. The increase of movement with smaller particle sizes was readily apparent, as was the increase in activity with temperature and the variation according to the nature of the liquid. Like Ramsay, Gouy excluded all exterior causes and immediately stated that Brownian movement actually evidences the internal agitation of a liquid, i.e. the movement of molecules,[118] and that it is a direct and visible proof of the modern hypothesis on the nature of heat.

In a more radical step, Gouy further postulated that the Brownian movement visibly demonstrates a fundamental and controversial problem in thermodynamics: that Brownian movement itself seems implicitly contrary to the absolutist interpretation of Carnot's principle. If, Gouy suggested, one suspended a solid Brownian particle by a very thin, frictionless filament wound about a wheel, the impulsions undergone by the particle, in its Brownian motion, would turn the wheel eternally, and one would achieve the irrealisable mechanism of perpetual motion. Perhaps, he suggested, Carnot's principle is in fact inapplicable to dimensions of very small magnitude, even of one micron, and applies only to larger, macromolecular mechanisms.[119]

In a more detailed experimental paper the following year (1889), Gouy refuted, point by point, all previous theories of Brownian movement other than the kinetic.[120] He also admonished observers not to confuse the behaviour of particles in suspension with those which adhere to the walls of the glass and warned that coagulation was quite different from the inhibition of Brownian movement. A few particles might remain isolated from the general coagulum and would persist in their irregular movement, despite the addition of salts, etc., which were merely coagulating agents. Gouy further cleared up the contention that Brownian movement occurred only in water, definitively establishing that it could be demonstrated with essentially the same characteristics in aqueous solutions, acids, alcohols, ethers, and hydrocarbons.

He demonstrated, too, that the movement persisted when all exterior agitation was eliminated (as testified by the unperturbed surface of a basin of mercury in the wee hours of the morning),[121] when an even temperature was positively insured (by a water bath), when illumination from different ranges of the spectrum was utilised, and when other circumstances of the environment were altered (including the introduction of an electromagnetic field). In answer to Nägeli's objection, Gouy initially cited the 'law of large numbers', without any elaboration of the subject, saying later that when one considers a large ensemble of particles and mean regularities, 'There is not any contradiction there, no more than with other phenomena governed by chance alone.'[122]

As mentioned earlier, the last of Gouy's major articles on Brownian movement appeared in the January, 1895, issue of the *Revue générale des sciences*, curiously and largely unnoticeably juxtaposed against Ostwald's article, 'La déroute de l'atomisme contemporain,' published in mid-November. Here Ostwald vehemently castigated the atomic, materialistic theory in that 'an immediate consequence of this theory has never been able to be verified!'[123] The kinetic theory, Ostwald accused, had yielded only continual contradiction between experimental facts and those results predicted under its auspices; its very foundations were wrecked by dependence upon the imaginary concepts of atoms and molecules, and chemists and physicists had been unnecessarily '. . . disturbing us with forces, the existence of which we cannot demonstrate, acting between atoms which we cannot see'.[124]

In contradistinction, Gouy unhesitatingly defended the role of hypothesis in general.

> It is necessary now to quit the solid terrain of observation and experience, in order to enter into the uncertain domain of hypotheses on the constitution of matter. These theories and hypotheses have been much abused, they have been much slandered; and yet we cannot do without

> them; their scientific importance is incontestable and they at times shed an unexpected light on a whole ensemble of questions. The history of the physical sciences shows us in effect that theoretical speculations have been the origin of the greatest progress and the most beautiful crop of discoveries. Let us accord to them that which is their due: the consideration which credits eminent services, and the limited confidence which never falls asleep and never neglects any means of control.[125]

In particular, Gouy spoke of the molecular hypothesis and its attendant kinetic theory of matter, and he unstintingly praised Brownian movement for its visible demonstration that what appears to be in repose to the naked eye is in reality in violent movement. Lauding the fruitfulness of kinetic theory, Gouy pointed out that it both predicts and explains the essential traits of the fifty-year-old experimental problem of Brownian movement. One may never see the movements of molecules (as Ostwald had rhetorically demanded), but one might view in Brownian movement the necessary result of molecular agitation. In conclusion, Gouy warned that the physicists' task had only just begun; the problem of the so-called molecular movements must be further studied by the methods of physics in order that the intimate properties of matter might better be known.

A New Approach in the Face of Idealism

Gouy's challenge in the *Revue générale des sciences* to his colleagues in physics was issued at a time when the general controversies over methodology and epistemology in the sciences were swelling to a high tide in France. If the mechanical approach to physical problems had been successful in the past, it was partly due to the long tradition intimately linking the Newtonian mechanical synthesis with a broad range of problems in the universe. With the Comtian interpretation of the value of science to society in the nineteenth century, certain difficulties arose as the traditional (mechanical) methods and results of the physical sciences were conscientiously and relentlessly further applied to the biological and social sciences.

The traditional mechanical method of science had in the past posed models of physical phenomena, sometimes choosing among alternatives and attaching the label of 'truth' or 'reality' to an especially successful model; it had nearly always suggested that an ultimate conception and realisation would finally be reached by persistent craftsmanship. Yet as the Comtians applied this method to the biological and social sciences, a profound pessimism set in, edging out the high hopes of the mid-century. The complexity and confusion ultimately resulted in Dubois-Reymond's 'Ignorabimus', rejecting the concept of ultimate knowledge of the origins of life, of the rules of conduct of life, and of man's destiny. This unexpected (and highly contested) result of the Comtian systemisation of the mechanical picture not only appeared to catapult impassable mysteries into the path of science, but also to countermand the associational psychology and concomitant methodology at its base. Rather than proceeding from the simple to the complex, by an association of ideas based upon the alliance of sense experience and Locke's innate sense perception,[126] one might just as well attempt to perceive immediately the general—the truth—a truth which traditional science had originally promised to bestow upon humanity and had henceforth failed to give.

There ensued in general, and perhaps most vociferously in France, what has been called the 'idealist reaction'. As normally used, the phrase refers to that *fin-de-siècle* loss of sympathy with science in general and an almost public nausea at the failure of science to cure social ills. It seemed that science had reduced man to a mechanical automaton and had dissected the forces of his environment into a blind determinism, describing them with cold precision in the annals of chemistry, physics, biology, and the new social sciences; and yet science had stopped far short of replacing man's broken graven images.

While in physics both the kinetic and energetic (thermodynamic) viewpoint suffered invective from the social and religious 'idealists', it was the overtly mechanistic conception that was most under fire. This occurred, too, at a time when the mechanists were undergoing the severe critique of the scientific 'idealists'—the so-called logical positivists who were concerned with direct sense perceptions and a strict economy of thought. As will be seen shortly, the denouement of this state of affairs involved in part a redefinition of the aims and methods of the physical sciences as well as experimental successes in more precisely defining the fundamental corpuscular hypothesis of matter. Those figures in France who helped effect this resolution—not least among them, Jean Perrin—were profoundly influenced by both idealist reactions.

The catchword of the social-idealist reaction in France was the 'bankruptcy' (*banqueroute*) of science, contrasting the promises of science to unfold all mysteries with its failure to shed light on what philosophers, *litterateurs,* theologians (and scientists, too) considered man's more important questions. The attack on science was launched in France by the literary historian Ferdinand Brunetière, who was then *maître de conférences* at the École normale, and who interpreted a play by Charles Bourget—*Le disciple* (1889)—most unfavourably to science, asserting that the hero's behaviour in the play was exemplary of the immorality of mechanistic scientific doctrines such as determinism and fatalism.[127] Brunetière then followed this initial blunderbuss with an article in the *Revue des deux mondes* in 1895 in which, playing upon Renan's statement in *L'avenir de la science* (1890), that the legitimate aim of modern science was to scientifically organise the world, he accused the physical and natural scientists of having failed to establish the Comtian positivistic religion which they had promised; Brunetière further warned that religion was now to have its turn.

From without the portals of science Brunetière was bolstered in France by the philosophers Félix Ravaisson, Émile Boutroux, Henri Bergson, Maurice Blondel, and Georges Fonsegrive, as well as by the literary figures Bourget, François de Curel, Eugëne Melchior de Vogüe, and Joris-Karl Huysmans.[128] Anatole France wrote concerning the intellectual crisis,

> It is most clear that confidence in science, which we have had so strongly, is more than half lost. . . . Ernest Renan . . . avows himself, without repudiating his faith, that there was some illusion in thinking that a society can be founded today entirely on rationalism and experience . . . youth is currently seeking something else.[129]

Although Alfred Fouillée cautioned that the current reaction against science need not jump to sceptical or mystical conclusions, and that it would be most useful if it called forth a new philosophy of action,[130] an avalanche of emotional replies quickly bombarded Brunetière, the idealists and mystics, with those authors writing in defence of science including the non-scientists Alfred Capus, Georges Clemenceau, Jules Soury, and, in the Chamber of Deputies, Jean Jaurès.[131]

The stampede of Brunetière and his cohorts into the halls of science evoked a variety of responses from French scientists themselves—responses which were closely linked with their opinions on the internal problems of chemistry and physics. The phenomenon which was occurring was indeed an international one, as was testified by the physicist Philipp Frank in a discussion of his student days at Vienna:

> At the turn of the century the decline of mechanistic physics was accompanied by a belief that the scientific method itself had failed to give us the 'truth about the universe'; hence nonscientific tendencies gained momentum. . . . A great many people believed that the time had come to return to the medieval ideas that may be characterised as the organismic conception of the world. . . . Two characteristic beliefs of nineteenth-century science broke down during its last decades . . . the belief that all phenomena in nature can be reduced to the laws of mechanics, and the belief that science will eventually reveal the 'truth' about the universe.[132]

Indeed, Ostwald's relentless opposition to mechanistic physics was partially in reaction to this very decay into seemingly irresoluble problems, such as the application of atoms to mental phenomena,[133] and the efforts of Ernst Mach were largely for the purpose of bolstering the bulwarks of science against metaphysical or spiritual incursions.[134]

In France, a chorus of voices offered different defences or accommodations to the phalanx of assembled critics. Some, like Berthelot, defended the traditional Comtian interpretation of science, while other scientists, such as Duhem and Poincaré, attempted an accommodation to criticism by a redefinition of the goals and methods of science. Poincaré, in particular, developed a philosophy of science in synthetic fashion, one of great influence upon the younger generation of physicists and chemists such as Jean Perrin, who was at this time completing his studies at the École normale.

The first scientist of renown to reply to the charge of 'banqueroute' had been Charles Richet, editor of the *Revue scientifique* and recipient of the Nobel prize in medicine in 1913 for his development of serum therapy and discovery of anaphylaxis.[135] Richet (1895) conceded that science had left certain mysteries unanswered, but denied that any promise had been broken; science had never taken the 'why' for its province, but only the 'how'. At the same time, science was linked with moral progress; morality, the goal of which was human solidarity, was the result of the intellectual development of man and the simultaneous growth of scientific knowledge.[136] Berthelot took essentially the same stance as Richet, placing the basis of morality in experience and knowledge, and asserting that through the scientific method of observation and experiment, man could advance his knowledge of the universe and of himself.[137] The certitude of this scientific method depended, as has been seen, upon experiment and upon restraint from exercises of the imagination, especially flights of the imagination into unverifiable hypotheses. Richet, too, avoided statements about the unseen or the essential nature of things. The chemist who analysed the molecular structure of a body, Richet remarked, knew quite well that his diagram was purely symbolic.[138]

In much more sophisticated fashion, the physicist Pierre Duhem replied to those critical of the methodology and goals of science with his own redefinition of physics and its theories. Duhem had become *professeur de physique théorique* at Bordeaux in 1895, after tenure both at Lille and Rennes. In addition to his mathematical studies of the mechanics of fluids and of thermodynamics and chemistry,[139] he had begun publishing in both the philosophy and the history of science. Though he remained at Bordeaux throughout the rest of his life, he hoped to be called to Paris, for he believed that a position in Paris would be the most effective means for his influencing the orientation of physicochemical researches.

Duhem's response to the two idealist reactions at the turn of the century was quite similar to that of Mach—a radical empiricist position which, unlike Mach, was somewhat Aristotelian in viewpoint and which seemed to echo the cry of Pascal: 'Humiliez-vous, raison impuissante!'[140] Duhem's principal faulting of traditional, mechanistic physical theory was that its object lay in the *explication* or *explanation* of a group of established experimental laws. 'Therefore, if the aim of physical theories is to explain experimental laws,' Duhem bewailed in his manifesto *La Théorie physique, son objet et sa structure* (1906), 'theoretical physics is not an autonomous science; it is subordinate to metaphysics.'[141] It is metaphysics, not science, which must deal with questions of reality—such as the problem of the existence of a material reality distinct from sensible appearances and the nature of this reality. The universal acceptance of any physical theory becomes absolutely impossible if individuals recognise in a theory a brand of metaphysics not their own.[142] Physical theory was, rather, to be purely descriptive and symbolic, though perhaps eventually becoming almost asymptotic to metaphysics as its own successes crescendoed.[143]

Regarding the use of hypothesis in physical theory, Duhem recognised two types—the model and the analogy. Duhem did not deny that mechanical models might have been profitable in the past nor that they were used by great physicists in the present; but Kelvin and Maxwell, he chided, were most brilliant when they slipped away from their models.[144] Too often, Duhem continued, discoveries are attributed to mechanical models—to the imaginative hypothesis touching upon molecules, their movements and collision, for example—when in fact, they were first due to other procedures, procedures forgotten after a model is subsequently set up.[145]

> The history of physics shows us that the search for analogies between two distinct categories of phenomena has perhaps been the surest and most fruitful method of all the procedures put in play in the construction of physical theories . . . not only does it bring a notable intellectual economy since it permits one to transfer immediately to one of the theories all the algebraic apparatus constructed for the other, but it also constitutes a method of discovery.[146]

To Duhem, the development of thermodynamics, especially in the sense of Mach, represented the very epitome of a scientific method of analogy and classification; thermodynamics was the apex of inductively derived abstract science.[147]

Implementing in practical fashion his conception of physical theory, a conception founded perhaps partly in his own preoccupations with thermodynamics and partly in his attempt to reconcile science with religion by redefining the bounds of physics and metaphysics, Duhem published the *Commentaires sur la thermodynamique* and the *Traité d'énergétique ou de thermodynamique générale*, where his stated purpose was '. . . to attempt to formulate the general laws which all physical properties obey, without supposing *a priori* that these properties should all be reducible to geometry and to local movement'.[148] Besides this work on thermodynamics, Duhem published several memoirs on electricity and magnetism, analysing electrodynamic energies from the law of Joule and laws of induction, and ignoring the current movement of modern electric researches toward corpuscular theories.[149]

He felt that corpuscular hypotheses and mechanical models had vitiated the success of the researches of Lorentz and J. J. Thomson—that the Zeeman effect suggested that profound modifications were necessary in Lorentz' theory and that the mechanical models of Thomson might very well be obscuring insight into what was really occurring in the ionisation of gases.[150]

Due to his preoccupation, then, with thermodynamics and to his general interpretation of scientific methodology, Duhem was thoroughly opposed both to mechanical models and to atomistics. Though part of his aversion to atomism may have derived from his pessimism about the convenience with which the actually known physical phenomena could be derived from a corpuscular hypothesis, nevertheless his position on the question of molecular and atomic reality was unambiguous; any speculation in this vein was pure metaphysics and outside the province of science.

The response of Henri Poincaré to the crisis in scientific epistemology and to the intrascientific controversy over atomism and kinetic theory was initially quite similar to that of Duhem, though Poincaré's viewpoint soon expanded into a more influential philosophy of science and one considerably more amenable to experimental innovation. During the height of the furore surrounding the idealists' charge of 'banqueroute'—at a time when Jean Perrin and the new generation of scientists, such as the Curies and Paul Langevin, were first undertaking researches—Poincaré was professor of mathematical physics at the Sorbonne (1887–1900) and a man of great prestige after his development of Fuchsian functions and work in celestial mechanics. His university lectures on optics, electricity, electromagnetic theory, thermodynamics and the calculus of probabilities were prepared for publication in a number of books, and his personal response to the excesses of Comtism, logical positivism and the idealist reaction were explored in three widely read books, *La science et l'hypothèse* (1902), *La valeur de la science* (1905), and *Science et méthode* (1909).[151]

Coming to grips with the current criticisms of science, Poincaré pinpointed the general problem:

> The ephemeral nature of scientific theories takes by surprise the man of the world. Their brief period of prosperity ended, he sees them abandoned one after another; he sees ruins piled upon ruins; he predicts that the theories in fashion today will in short a time succumb in their turn, and he concludes that they are absolutely in vain. This is what he calls the *bankruptcy of science*. . . .[152]

Yet, asserted Poincaré, what decays in theories is that which claims to teach what things *are*; if an adjunct of the theory reveals, instead, a true relation between things, this *relation* is acquired for all time.[153] The equations of science express relations, and if the equations remain true, it is because the relations preserve their reality. For example, 'they teach us now, as they did then, that there is such and such a relation between this thing and that; only the something which we then called *motion*, we now call *electric current*';[154] the real objects for which man substitutes images and names will remain hidden forever from man's eyes.[155]

Within the scientific community, Poincaré avoided a stance associating him with either the energetic or the mechanistic physics.[156] In keeping with his general philosophy of science, he criticised the classical mechanical treatises for their failure to distinguish clearly between what is experiment, what is mathematical reasoning, what is convention, and what is hypothesis. 'The principles of dynamics appeared to us first as experimental truths,' he wrote, 'but we have been compelled to use them as definitions. . . . [E]xperiment may serve as a basis for the principles of mechanics, and yet will never invalidate them. . . . The important thing is not to know what force is, but how to measure it.[157] The relationship of force to mass and acceleration is beyond the reach of any future experiment, he wrote, and the fallacy of mechanical theories involving images and models, rather than concentrating on functional relationships, is easily seen for, 'If a phenomenon is susceptible of one mechanical explanation, it is susceptible of an infinitude of others which would account equally well for all the features revealed by experience.'[158]

Regarding the role of hypothesis in scientific theory, Poincaré distinguished three types of hypothesis. Those which are natural and necessary—as that small movements obey a linear law, and that effect is a continuous function of its cause—are the common basis of all the theories of mathematical physics. They are the last that should be abandoned. A second category of hypotheses are indifferent: whether matter is continuous or is formed of atoms, the results would be the same. On the atomic supposition there is a little more difficulty in obtaining results—that is all. The hypotheses of the third category are real generalisations and must be confirmed or invalidated by experiment. Whether verified or condemned, they will always be fruitful; but they will only be so if they are not too numerous.[159]

The role of the atom as hypothesis, and in particular, its utilisation in the kinetic theory of gases was a specific problem with which Poincaré grappled throughout his life. In 1894, in a paper disputing with Lord Kelvin a particular point in the kinetic theory, Poincaré questioned its merits, concluding that it could not account for all the known experimental facts, that it could neither be deemed true nor false, and that the question of its sterility or fecundity as an aid in discovery remained wide open. Crookes, he cited, had found it useful in his work on radiant matter; it had also aided in revealing the relation between gaseous and osmotic pressures.[160] In *Science and Hypothesis* (1902), Poincaré stated substantially the same position.

In his first lectures on mathematical physics, published 1887–1889, he prefaced his treatment of thermodynamics by lauding the avoidance of molecular hypothesis: 'Abandoning the ambitious theories of forty years ago, encumbered with molecular hypotheses,' he wrote, 'we search today to raise on Thermodynamics alone the entire edifice of mathematical Physics', and praising the simplicity of the first law, he affirmed that 'we refuse to ourselves the belief that an unexpected harmony should be a simple effect of chance'.[161] This is not to say, however, that he was an energeticist, for he explicitly posed the law of the conservation of energy as encompassing only the statement that 'There is something which remains constant.'[162]

Though Poincaré was thoroughly familiar with Boltzmann's interpretation of the kinetic theory (it was he who delivered the *éloge* to Boltzmann at the Académie des Sciences in 1906),[163] he questioned profoundly its mathematical rigour and was, with others, bothered by the incongruity of reversible molecular collisions with the irreversibility of nature's processes and of Carnot's principle. 'A strictly mechanical explanation of these phenomena has . . . been sought', he wrote, 'but, owing to their nature, it is hardly likely that it will be found.' The kinetic theory attributes the confusion to the crudity of our senses, and,

> The apparent irreversibility is therefore but an effect of the law of great numbers. Only a being of infinitely subtle senses, such as Maxwell's demon, could unravel this tangled skein and turn back the course of the universe. This conception . . . has cost great effort and has not, on the whole, been fruitful. . . .[164]

Yet, unlike so many of his colleagues who also had reservations about the kinetic-molecular conception, Poincaré was aware of the new possibility to experimentally verify the theory, and in 1902 he called to the public's notice '. . . the original ideas of M. Gouy on the Brownian movement. According to this scientist, this singular movement does not obey Carnot's principle. The particles which it sets moving would be smaller than the meshes of that tightly drawn net. . . . One can almost see Maxwell's demon at work.'[165] Again, at the Congress of Arts and Science at St. Louis in 1904, he mentioned Gouy's surmises on the transformation of Brownian movement into heat by friction, the inverse transformation of heat into motion, and the apparent violation of Carnot's principle since the motion lasts forever. 'If this be so,' Poincaré mused, 'we no longer have need of the infinitely subtle eye of Maxwell's demon; our microscope suffices us.'[166]

Yet, questioned Poincaré, if the violations of Carnot's principle are encountered only among infinitesimals—under the microscope or with electrons and radioactivity—perhaps another infinitesimal remains yet unseen and counterpoised to the first. Only experiment could solve the question, and whilst the experimentalists determine the question, physics must proceed as if the principles remained uncontested. With remarkable insight into the future of physics and a bit of conservatism, too, Poincaré concluded his 1904 speech by remarking that

> We cannot foresee in what way we are about to expand; perhaps it is the kinetic theory of gases which is about to undergo development and serve as a model to the others. . . . Physical law will then take an entirely new aspect; it will no longer be solely a differential equation; it will take the character of a statistical law. Perhaps likewise, we could construct a whole new mechanics. . . . The ordinary mechanics, more simple, would remain a first approximation, since it would be true for velocities not too great, so that we still find the old dynamics under the new . . . I hasten to say in conclusion we are not yet there, and as yet nothing proves that the principles will not come forth from the combat victorious and intact.[167]

The Question of Molecular Reality becomes Acute

In addition to these profound analyses of Duhem and Poincaré, there was in this period at the turn of the century a barrage of other articles in the scientific literature discussing the fecundity of mechanical models, atoms and molecules, and the kinetic theory. Crookes' articles were translated, as well as those of Sir Arthur Rücker and other Englishmen, and the new experimentation on cathode rays and radioactivity was excitedly portrayed to the readers of the *Revue scientifique*, *Revue des questions scientifiques*, *Revue générale des sciences*, and other periodicals which touched upon science; in France, as elsewhere, the *Comptes rendus*, *Journal de physique*, and other scientific journals recorded innumerable experiments with these new phenomena. With controversy rampant on both the professional and public levels, the market was ripe for the publication of books and pamphlets, and a number of historical and philosophical treatments quickly appeared with diverse axes to grind. Leopold Mabilleau, for example, published a *Histoire de la philosophie atomistique* in 1895, while in his *Essai critique sur l'hypothèse des atomes* (1895), Arthur Hannequin handled the atom as a concept, not a thing in itself, insisting that its contradictions could be resolved only by metaphysics.[168] In a more widely received book, Émile Picard espoused essentially the philosophic position of Poincaré, but with a more positive emphasis upon the utility of the molecular-kinetic hypothesis. Picard wrote that although, as Poincaré had pointed out, an infinite number of mechanical models might be possible to explain physical phenomena, history had shown again and again a tendency to return to the atomistic and molecular conceptions, and even today, the discontinuous hypothesis was assuming a new vitality.[169] Agreeing with Poincaré that it was most valuable for science to utilise diverse points of view in order to progress, Picard himself displayed great enthusiasm for the mechanical viewpoint, and like Poincaré, he called attention to the experiments of Gouy on Brownian movement as a challenge to the traditional energetic interpretations of Carnot's principle.

A more thorough appraisal of the crisis in physics was undertaken by Abel Rey in 1907 in his *thèse*, *L'énergétique et le mécanisme au point de vue des conditions de la connaissance*,[170] in which he described science as a product of psychological and social factors. Mechanics, he wrote, owed a large part of its success to its ease of popularisation, since science can only be developed in contact with an entire society and a society which is interested in its results. Energetics, Rey showed, had tried to popularise its tenets, but had found this impossible since the point of departure was the differential equation itself. The pure concept which the energeticists tried to publicise was only a myth, and a dangerous one at that. The energeticist approach, Rey asserted, concerned only natural laws which one already knew, observations already made, and phenomena already studied. The energetic viewpoint could never propose any means of discovery; 'Like the demiurge of Greek conceptions, the energetics theory organises but does not create.'[171]

Within the scientific community itself, few French scientists were willing to envision the wholesale revamping of physics suggested by Poincaré and implied in the papers of Léon Gouy, which Poincaré and Picard had cited, nor did they necessarily agree with Rey's analysis of the poverty of energetics and thermodynamics. As William Crookes remarked in an article published in the *Revue scientifique*, 'This dream [of atoms and molecules] is essentially a British dream, and the audacity with which we have thrown ourselves into speculations and hypotheses has almost given the lie to our traditional reputation as an exclusively practical people. We have discarded the notion of impenetrable mysteries.'[172] Indeed, mysteries remained on the Continent, as the viewpoint of science as a description of phenomena by means of differential equations continued to have great currency. Mirroring the discord and confusion was a remark in a review of Mach's *Mechanics* that '. . . for the moment at least, this [Mach's] point of view tends to predominate among physicists; nevertheless molecular and atomic representations and the old, somewhat anthropomorphic notions of traditional mechanists have, no doubt, not yet exhausted their fertility.'[173]

One of the more adamant among the opponents to the molecular-kinetic viewpoint at this time was Henry Le Chatelier, who was *professeur de chimie industrielle* at the École des Mines from 1887 to 1897 and then *professeur de chimie minerale* at the Collège de France. Le Chatelier was known for his studies of chemical equilibrium, the combustion of gas mixtures, and the completion of the law of phases of Gibbs. In assessing the aims of science, he insisted that practical applications must be a primary goal, rather than theories and notations as things in themselves; that knowledge of nature was to be sought 'such as it really exists, and not such as it could please us to imagine it'.[174] Atomic notation and formulas, he assessed, had been of great use to chemistry, but once they became associated with atomic theory, the result was only tremendous discord among chemists, to the detriment of scientific progress. The task of chemists and physicists lay in following the path traced by their *maîtres*, honouring that French science which limits its investigations to real facts and to the laws which group them.[175]

Though he disagreed with Ostwald's more extreme pronouncements on energy,[176] Le Chatelier supported the 1895 'Déroute de l'atomisme contemporain', as did other scientists, including G. Charpy, A. Étard, and Philipp Guye.[177] This 'déroute', however, excited the ire of the few proponents of the kinetic theory in France, and an immediate reply proceeded from the pen of Marcel Brillouin, *maître* of the course of physics at the École normale and the major professor of Jean Perrin. Indeed, Brillouin was among the few professors who taught the kinetic theory from a thoroughly sympathetic viewpoint at this time.[178]

In his argument 'Pour la matière', Brillouin attacked specific weak points in Ostwald's discussion, such as the purported non-addivity of chemical properties and demise of the ondulatory theory of light. Regarding the kinetic theory, Brillouin conceded that it did present certain difficulties, but that 'one should admire what marvellous advantage Clausius has been able to draw from a unique notion—the inevitable consequence of the spontaneous diffusion of a gas despite its weight—that of the peculiar movement of the constituent parts of the gas.'[179] As to the merits of mechanical models in scientific methodology, Brillouin insisted upon the right of employing mechanical images without being excommunicated from the scientific community, or being accused of backwardness; further, he applauded the simultaneous usage of both the differential equations of energetics and the intuitive models of the kinetic theory.[180]

To this plea, Ostwald replied,

> For six years I have searched without success to construct a mechanical theory of chemical affinities, and I am convinced that it is only as one has renounced all mechanical analogy that one can find results of some utility. It is clear that my adversary has not arrived at a similar impression. . . .[181]

Indeed, not only was Brillouin to remain immovable in his support of the kinetic theory, but he was also to convert one of his most apt pupils Jean Perrin, to his way of thinking. Whereas Brillouin, the mathematical physicist, failed to sway Ostwald from his unyielding opposition to the molecular-atomic hypothesis and its concomitant kinetic theory, the pupil and experimentalist, Perrin, was to wage a victorious campaign against the radical empiricist and positivistic camp, forcing a general acceptance of molecular reality and a recantation by even the high priest of energetics—Ostwald.

NOTES

1. Maurice Caullery, *La Science française depuis le xviie siècle* (Paris: Colin, 1948), p. 129.
2. Ibid., p. 172.
3. See T. E. Thorpe, *Essays in Historical Chemistry* (London: Macmillan, 1894), pp. 366–381.
4. Harold C. Urey, 'Dalton's Influence on Chemistry', pp. 329–343 in *John Dalton and the progress of science*, ed. D. S. L. Cardwell (New York: Manchester University Press, 1968), p. 333. Dalton never accepted Gay-Lussac's law or Avogadro's hypothesis. See also, in the same volume, Wm. H. Brock, 'Dalton versus Prout: The Problem of Prout's Hypothesis' (pp. 240–258), pp. 246–247, 255.

5. Brock, pp. 248–250, discusses the tables for atomic weights of Berzelius and of Thomas Thomson. Berzelius' figures were fractional, whereas Thomson's figures conformed by and large to the integral multiples law of Prout. Although Thomson's values were widely accepted by 1827, his methods and precision were severely criticised in subsequent years by leading chemists, including Dalton, Berzelius, Turner, and Penny (254).
6. William Whewell, *The Philosophy of the Inductive Sciences* (London: John W. Parker, 1847), Vol. I, p. 424.
7. Humphry Davy, 'Discourses of the President, Anniversary, 1826', pp. 92–99 in the *Collected Works*, Vol. 7 (London: Smith, Elder & Co., 1840), pp. 96–97.
8. Brock, p. 248, terms Wollaston's distinction between experimentally determined equivalent weights and hypothetical atomic weights 'positivist' and writes that it was puzzling to Dalton.
9. Stanislao Cannizzaro, 'Considerations on some points of the theoretic teaching of chemistry', pp. 17–33 in *Faraday Lectures 1869–1928*, ed. C. S. Gibson and A. J. Greenaway (London: The Chemical Society, 1928), p. 21.
10. See Brock.
11. C. A. Russell, 'Berzelius and the Development of the Atomic Theory', pp. 259–273 in Cardwell, p. 263.
12. Ibid., p. 262.
13. Ibid., p. 268. Quoted from Berzelius in the *Annals of Philosophy*, **2** (1813), p. 450.
14. Cannizzaro, p. 20.
15. Ibid.
16. Thorpe, pp. 366–381.
17. See W. H. Brock and D. M. Knight, 'The Atomic Debates', *Isis*, **56** (1965), pp. 5–25; in *The Atomic Debates* (Leicester University Press, 1967), pp. 1–30.
18. Quoted from E. Cohen, *Vijftig jaren revolutie* (Amsterdam, 1939), p. 10 in Andrew G. van Melsen, *From Atmos to Atom* (New York; Harper Torchbooks, 1960), p. 151.
19. Quoted from Kekulé (1867) in Ida Freund, *The Study of Chemical Composition* (Cambridge: Cambridge University Press, 1904), p. 624.
20. M. P. Crosland, 'The First Reception of Dalton's Atomic Theory in France', pp. 274–287 in Cardwell, p. 274.
21. Ibid., p. 275.
22. Ibid., p. 276.
23. Ibid., pp. 279–280.
24. Ibid., p. 278.
25. Brock, p. 247.
26. René Taton, *Causalités et Accidents de la Découverte Scientifique* (Paris: Masson et Cie, 1955), p. 141.
27. See Thorpe's chapter on Dumas in the *Essays in Historical Chemistry*.
28. Aaron J. Ihde, *The Development of Modern Chemistry* (New York: Harper and Row, 1964), p. 196.
29. Caullery, p. 125.
30. Brock, p. 247.

31. Ibid., p. 153.
32. Ibid., p. 128.
33. Ibid., pp. 228–230.
34. M. A. Haller, 'Inauguration de la Statue de Adolphe Wurtz', *Institut de France. Académie des Sciences* (Paris: Gauthier-Villars, 1921), 11 pp.
35. For a discussion of the Wurtz-Berthelot polemics, see the article by Reino Virtanen, 'Marcelin Berthelot. A Study of a Scientist's Public Role', *University of Nebraska Studies*, new series #31 (April, 1965), 67 pp. See also the summary of the polemics (in *Comptes rendus*, **84** (1877), pp. 1110, 1183, 1189, 1191, 1194, 1264 f., and 1269 f.) entitled 'Atoms and Equivalents', in *Nature*, **16** (1877), pp. 293–294 and signed J. M. T. (probably John M. Thomson).
36. Charles Adolphe Wurtz, *La Théorie Atomique* (Paris: Librairie Germer Baillière et Cie, 1879), p. 2.
37. Marcelin Berthelot, *Les Origines de L'Alchimie* (Paris: Georges Steinheil, 1885), p. 291.
38. Caullery, p. 129.
39. Paul Painlevé, 'La philosophie de Marcelin Berthelot', *Revue du Mois*, **3** (1907), p. 514.
40. Virtanen, p. 4.
41. Berthelot, *L'Origines de L'Alchimie*, p. 290.
42. Robert Brown, 'A brief account of microscopical observations made in the months of June, July, and August, 1827, on the particles contained in the pollen of plants; and on the general existence of active molecules in organic and inorganic bodies', *Philosophical Magazine*, **4** (1828), p. 171.
43. Thorpe, p. 266.
44. Stephen Gray, for example, published a paper in the *Philosophical Transactions*, **19** (1696), pp. 280–287, describing 'some of those Minute Insects, which the deservedly famous Mr. Leewenhoek Discovered by the help of Excellent Microscopes: . . . I saw them and some other Irregular Particles. . . . They have a two-fold motion, a swift progressive irregular one and at the same time a rotation on their Axes at right angles to the diameter' (pp. 282–283).
45. Adolphe Brongniart, 'Mémoire sur la génération et le développement de l'embryon dans les végétaux phanerogames', *Annales des Sciences Naturelles*, **12** (1827), p. 41.
46. Ibid., p. 44.
47. Ibid., p. 45.
48. A translation of Raspail's article was published a year later in the *Edinburgh Journal of Science*, **10** (1829), pp. 96–106, entitled 'Observations and Experiments tending to demonstrate that the Granules which are discharged in the explosion of a grain of Pollen, instead of being analogous to spermatic Animalcules, are not even organized Bodies'. See p. 99.
49. Adolphe Brongniart, 'Nouvelles recherches sur le pollen et les granules spermatiques des végétaux', *Annales des Sciences Naturelles*, **15** (1828), pp. 381–390.
50. Brown, 'A brief account . . .', pp. 162–163.
51. Ibid., pp. 164–165.
52. Ibid., pp. 165–166.

53. Brongniart, 'Nouvelles Recherches . . . . Note Additionelle', p. 394.
54. Ibid.
55. Ibid., p. 396.
56. Brown, 'Additional Remarks on Active Molecules', *Edinburgh New Philosophical Journal*, **8** (1830), pp. 43–44.
57. M. Raspail, 'Note on Mr. Brown's Microscopical Observations on the Active Molecules of Organic and Inorganic Bodies', *Edinburgh New Philosophical Journal*, **8** (1830), pp. 43–44.
58. Hrn. Hofr. Muncke, 'Ueber Robert Brown's mikroskopische Beobachtungen, über den Gefrierpunkt des absoluten Alkohols, und über eine sonderbare Erscheinung an der Coulomb'—schen Drehwaage', Poggendorf's *Annalen der Physik*, **17** (1829), pp. 159–165.
59. David Brewster, 'Observations Relative to the Motions of the Molecules of Bodies', *Edinburgh Journal of Science*, **10** (1829), p. 219.
60. Karl Michael Marx, 'Sur les mouvemens des molécules', *Bibliothèque universelle des Sciences, Belles-Lettres, et Arts de Genève*, **48** (1831), p. 121.
61. Félix Dujardin, *Nouveau Manuel Complet de L'Observateur au Microscope* (Paris: Roret, 1843), p. 60.
62. Cannizzaro, p. 26.
63. Wurtz, *La Théorie Atomique*, p. 239.
64. Thorpe, p. 370.
65. René Dugas, *La Théorie Physique au Sens de Boltzmann et ses Prolongements Modernes* (Paris: Dunod, 1959), pp. 78–80.
66. Robert S. Cohen, 'Hertz's Philosophy of Science: An Introductory Essay', in Heinrich Hertz' *The Principles of Mechanics*, trans. D. E. Jones and J. Walley (New York: Dover, 1956), sections 2 and 3.
67. F. G. Donnan, 'Ostwald Memorial Lecture', *Journal of the Chemical Society*, Part I (1933), p. 327.
68. Wilhelm Ostwald, *Die Energie*, 2nd ed. (Leipzig: J. A. Barth, 1912), pp. 126–127.
69. *Aus dem wissenschaftlichen Briefwechsel Wilhelm Ostwalds*, ed. Hans-Günther Körber and Grete Ostwald (Berlin: Akademie-Verlag, 1961), pp. xvi–xvii.
70. Wilhelm Ostwald, 'La Déroute de L'Atomisme contemporain', *Revue generale des sciences*, **6** (1895), p. 956.
71. Wilhelm Ostwald, 'Elements and Compounds', p. 187 in *Faraday Lectures 1869–1928*. From the definition that 'a substance or a chemical individual is a body, which can form hylotropic phases within a finite range of temperature and pressure,' Ostwald deduced the laws of constant proportions, multiple proportions, and combining weights in his Faraday Lecture of 1904. As he remarks, p. 196, no one else had mentioned such a deduction without the help of the atomic hypothesis except to deny the possibility.
72. Quoted from E. Cohen in van Melsen, p. 151.
73. Abel Rey, *L'Énergetique et le mécanisme* (Paris: Félix Alcan, 1907), p. 91. Rey bases his viewpoint on van't Hoff's *Leçons de Chimie Physique*.
74. Erwin N. Hiebert, 'The Conception of Thermodynamics in the Scientific Thought of Mach and Planck', *Ernst Mach Institut. Wissenschaftlicher Bericht*, **5** (1968), pp. 24, 28.

75. Ibid., p. 50.
76. Max Planck, *Vorlesungen über Thermodynamik* (Leipzig: Verlag von Veit & Co., 1897), p. iv. Curiously, the English translator omits the phrase 'of the accepted hypotheses'. See 'From the Preface to the 1st Edition', in Planck, *Treatise on Thermodynamics*, 3rd ed., trans. Alexander Ogg (London: Longman, Green & Co., 1927), p. viii.
77. Planck, *Treatise on Thermodynamics*, p. ix.
78. Carlton Berenda Weinberg, *Mach's Empirio-Pragmatism in Physical Science* (New York: Albec Press, 1937), p. 51.
79. Philipp Frank, *Modern Science and its Philosophy* (Cambridge, Mass.: Harvard University Press, 1949), p. 18.
80. Weinberg, pp. 78–79, 96.
81. Ernst Mach, *Erkenntnis und Irrtum* (Leipzig: J. A. Barth, 1905); see the chapter on hypothesis, pp. 229–246. Also quoted in Weinberg, p. 102.
82. Ernst Mach, *The Science of Mechanics*, 6th ed., trans. Thomas J. McCormack (La Salle, Illinois: Open Court Publishing Co., 1960), p. 599.
83. Weinberg, pp. 47, 100, 104.
84. Ibid., p. 97.
85. Boltzmann, 'Nochmals über die Atomistik', pp. 158–161 in the *Populäre Schriften*, ed. J. A. Barth (Leipzig, 1905), p. 160. Also in the *Annalen der Physik*, **61** (1897), p. 790; quoted in Dugas, p. 26.
86. Boltzmann, Über die Unentbehrlichkeit der Atomistik in der Naturwissenschaft', pp. 141–157 in the *Populäre Schriften*, pp. 144–145. Also in the *Annalen der Physik*, **60** (1897), p. 231; quoted in Dugas, pp. 103–104.
87. Dugas, p. 62.
88. Boltzmann, 'Über die Bedeutung von Theorien', (Graz, 16 July, 1890), pp. 76–80 in the *Populäre Schriften*, p. 77. Also quoted in Dugas, p. 122.
89. Boltzmann, *Vorlesungen über die Principe der Mechanik*, 2 vols. (Leipzig: J. A. Barth, 1897–1904), Vol. I, p. 6. Also quoted in Dugas, p. 62.
90. For example, Ostwald had contended that kinetic theory could not logically explain the irreversibility of paths followed by natural phenomena. Boltzmann, on the contrary, maintained that one of the most brilliant confirmations of kinetics was precisely in its excellent image of dissipation of energy, if one allowed the postulate that the world has strayed from an initial state corresponding to given conditions which may be called non-probable. See Boltzmann's 'Ein Wort der Mathematik an die Energetik', pp. 104–136 in the *Populäre Schriften*, p. 134. Also in the *Annalen der Physik*, **57** (1896), p. 39. Quoted in Dugas, p. 97.
91. Boltzmann, 'Über die Entwicklung der Methoden der theoretischen Physik in in neuerer Zeit', (Munich, 22 September 1899), pp. 198–227 in the *Populäre Schriften*, pp. 224–225. Also cited in Dugas, pp. 115–117.
92. Unpublished letter from Léon Gouy to Jean Perrin, dated from Lyon, the 31st of October, 1908.
93. Quoted from the *Philosophical Magazine*, 1828, in the *Royal Microscopical Society Journal*, **10** (1890), p. 120. Conybeare's comment was expressed in a manuscript letter inserted at the beginning of Brown's 1828 article and addressed to the Reverend Buckland, Oxford.

94. Léon Gouy, 'Sur le mouvement brownien', *Comptes rendus*, **109** (1889), p. 102. Gouy fails to cite the source for Maxwell's viewpoint.
95. Ch. Wiener, 'Erklärung des atomischen Wesens des tropfbarflüssigen Körperzustandes, und Bestätigung desselben durch die sogenannten Molekularbewegungen'. Poggendorf's *Annalen der Physik*, **118** (1863), pp. 79–94.
96. Ibid., p. 90.
97. Wiener's theory is a curious netherland between the older static conception of molecules and the newer kinetic interpretation. Prior to the mid-nineteenth century, it was assumed molecules reacted upon one another primarily through the 'action-at-a-distance' formulation, involving attractions and repulsions. In Wiener's conception, aether atoms lay in the spaces between matter atoms, repelling them, while the matter atoms were attracted to each other. This network of forces insured a system of eternal agitation.
98. According to Budde in 'Untersuchungen über die Brownsche Molekülarbewegung', *Verhandlungen des Naturhistorischen Vereines der preussischen Rheinlande und Westphahlens*, **27** (1870), p. 109, Binz studied six preparations of Brownian movement with water and different narcotics, including morphine, strychnine, and quinine, and demonstrated different rates of attrition of the Brownian movement, according to the narcotic used.
99. See, e.g. R. Meade Bache, 'The Secret of the Brownian Movements', *Proceedings of the American Philosophical Society*, **33** (1894), pp. 163–177. After attaining unsatisfactory results from such a series of studies, Bache concluded that Brownian movement was a property of water alone, due to the mutual repulsion of water molecules.
100. 'Brownian Movement in Milk', *New York Microscopical Society Journal*, **2** (1886), p. 101.
101. Sigmund Exner, 'Untersuchungen über Brown's Molecularbewegung', *Sitzungsberichte der kaiserlichen Akademie der Wissenschaften, mathematisch-naturwissenschaftliche Klasse, Wien*, **56** (1867), pp. 116–123.
102. J. B. Dancer, 'Remarks on Molecular Activity as shown under the Microscope', *Proceedings of the Manchester Literary and Philosophical Society*, **7** (1868), pp. 162–165.
103. Minerologists had long observed liquid inclusions in certain minerals. The geologist, de Lapparent, later discussed in his *Traité de Geologie* (1883) small bubbles, of dimensions less than 0·002 mm, in a state of constant trepidation similar to the movements of solid corpuscles in Brownian movement. These bubbles were found in minute liquid cavities within quartz and other substances.
104. Walter N. Hartley, 'An explanation of the "Brownian Movement"', *Monthly Microscopical Journal*, **18** (1877), pp. 8–9 and 'On attraction and repulsion of bubbles by heat', *Proceedings of the Royal Society*, **26** (1877), pp. 137–149.
105. G. Quincke, 'Ueber die Bewegung und Anordnung kleiner Teilchen, welche in Flüssigkeiten schweben', *Verhandlungen der Gesellschaft deutscher Naturforscher und Ärtze*, **2** (I) (1898), p. 29.
106. W. Stanley Jevons, 'On the so-called Molecular Movements of Microscopic

Particles', *Proceedings of the Manchester Literary and Philosophical Society*, **9** (1869–1870), pp. 78–84.

107. Maltezos' articles in the *Annales de Chimie et de Physique* (April, 1894) and the *Comptes rendus*, **121** (1895), pp. 303–305, are summarised in the *Royal Microscopical Journal*, **15** (1895), p. 591.
108. At a meeting of the New York Microscopical Society in 1889, the president informed the group that his slide of gamboge in water, with which he had displayed Brownian movement since August 3, 1874 had sprung a leak and ceased activity. These fourteen years, he said, had probably been the 'longest period during which this phenomenon has been under observation' (*New York Microscopical Society Journal*, **5** (1889), p. 46). His claim was apparently in error, however, as Charles Philippe Robin wrote in his *Traité du Microscope* (Paris, 1871), p. 528, that he still retained preparations of carbon and other granules in water, in each of which the movements had never ceased. These preparations, he remarked, were made in 1853.
109. Joseph Delsaulx, 'Thermodynamic Origin of the Brownian Motions', *Monthly Microscopical Journal*, **18** (1877), pp. 1–7.
110. Ibid., p. 2.
111. Ibid., p. 6.
112. J. Thirion, 'Les Mouvements Moléculaires', *Revue des questions scientifiques*, **7** (1880), p. 43. He also credits an unpublished note of R. P. Carbonelle (1874) with this viewpoint, antedating Thirion's own interest in the problem.
113. Ibid., p. 44.
114. Ibid., p. 51.
115. Von Nägeli, 'Ueber die Bewegungen kleinster Körperchen,' *Sitzungsberichte de mathematisch-physikalischen Classe der k.b. Akademie der Wissenschaften zu München*, **9** (1879), pp. 389–453.
116. From a summary of Ramsay's article in the *Proceedings* of the *Bristol Naturalists' Society*, **3** (1882), pp. 299–302, in the *Royal Microscopical Society Journal*, Ser. 2, **3** (1883), p. 612.
117. Wm. Ramsay, 'Pedetic Motion in Relation to Colloidal Solutions', *Chemical News*, **65** (19 February 1892), p. 90. The term 'pedesis' was proposed by Jevons, in lieu of 'Brownian motion' or 'molecular movement', in the *Quarterly Journal of Science*, 1878.
118. Léon Gouy, 'Note sur le mouvement brownien', *Journal de Physique*, **7** (1888), p. 563.
119. Ibid., p. 564.
120. Léon Gouy, 'Sur le mouvement brownien', *Comptes rendus*, **109** (1889), pp. 102–105.
121. Budde had preceded him in this demonstration. See Budde, p. 110.
122. Léon Gouy, 'Le Mouvement brownien, et les mouvement moléculaires', *Revue générale des Sciences*, **6** (1895), pp. 1–7. Also published at Lyon by A. Storck, 1895, in 23 pp.
123. Wilhelm Ostwald, 'La Déroute de L'Atomisme contemporain', p. 955.
124. Ibid., p. 957.
125. Gouy (1895), p. 14.

126. Charles Coulston Gillispie, 'Science and Enlightenment', in *The Edge of Objectivity* (Princeton University Press, 1960), p. 162. See the analysis by Abel Rey in *L'Énergétique et le mécanisme*, pp. 16–17.
127. See Virtanen's article and Harry Paul, 'The Debate over the Bankruptcy of Science of 1895,' *French Historical Studies*, **5**, #3 (1968), pp. 299–327 for a discussion of the 'banqueroute' polemics.
128. Paul, p. 300.
129. Quoted from France, *La Vie Litteraire*, **IV** (Paris, 1900), in Virtanen, p. 30.
130. From Alfred Fouillée's 'L'Hégémonie de la science et de la philosophie', *Revue philosophique*, **41** (1896), cited in Virtanen, p. 42.
131. Virtanen, pp. 34–35.
132. Frank, pp. 2–4.
133. *Aus dem wissenschaftlichen Briefwechsel Wilhelm Ostwalds*, pp. 140–141.
134. Frank, pp. 84–85.
135. Paul, p. 306.
136. Ibid., pp. 306–309.
137. Virtanen, pp. 36, 37.
138. Paul, p. 307.
139. For a short discussion of his scientific and other work, see Émile Picard, *La Vie et l'Oeuvre de Pierre Duhem* (Paris: Gauthier-Villars, 1922).
140. Virtanen characterises the French idealistic reaction as an echo of Pascal (pp. 29–30). In his analysis of Duhem, Picard dismissed the appellation of 'Kantian' for Duhem, explaining that although Duhem did insist upon physical theory as something purely formal, he regarded the *Critique of pure reason* as long, obscure, confused and pedantic (p. 50). Picard insists that it is Pascal whom Duhem constantly cites and Pascal's *Pensées* which Duhem knew by heart (pp. 50–51).
141. Pierre Duhem, *The Aim and Structure of Physical Theory*, trans. from the 2nd Fr. ed. Philip Wiener (Princeton: Princeton University Press, 1954), p. 10.
142. Ibid., p. 25.
143. Picard, p. 28. See also Duhem's *Le système du monde, histoire des doctrines cosmogoniques de Platon à Copernic*, Vol. I. (Paris: Hermann, 1913), pp. 134–150.
144. Duhem, *The Aim and Structure of Physical Theory*,pp. 26–27.
145. Duhem supports this contention with the development of the theory of continuity of the liquid and gaseous states and the theory of osmotic pressures as cases in point. These, he claims, were generalised from induction, using the methods of thermodynamics. See *The Aim and Structure of Physical Theory*, pp. 98–99.
146. Ibid., pp. 95–97.
147. Ibid.; see especially pp. 51–54.
148. Duhem, *Traité d'énergétique* (2 vols; Paris: Gauthier-Villars, 1911), Vol. I., p. 2.
149. Picard, p. 23.
150. Duhem, *The Aim and Structure of Physical Theory*, pp. 98–99.
151. All of these works exist in English translation: *Science and Hypothesis*, trans. W.J.G. (New York: Dover, 1952); *The Value of Science*, trans. George Halsted (New York: Science Press, 1907); and *Science and Method*, trans. Francis Maitland (New York: Dover, 1952).

152. Poincaré, *Science and Hypothesis*, p. 160.
153. Poincaré, *The Value of Science*, p. 139.
154. Poincaré, *Science and Hypothesis*, p. 161.
155. Ibid.
156. Abel Rey, *La Philosophie Moderne* (Paris: Flammarion, 1908), p. 137.
157. Poincaré, *Science and Hypothesis*, pp. 104–106.
158. Quoted from Poincaré's *Electricity and Optics* in Tobias Dantzig, *Henri Poincaré, Critic of Crisis* (New York: Charles Scribner's Sons, 1954), p. 10. Dantzig was a student of Poincaré at Paris.
159. Poincaré, *Science and Hypothesis*, pp. 152–153.
160. Poincaré, 'Sur la théorie cinétique des gaz', *Revue générale des sciences*, **5** (1894), p. 513.
161. Poincaré, *Thermodynamique. Cours de physique mathématique*, 2nd ed. (Paris: Gauthier-Villars, 1908), pp. v, viii.
162. Ibid., p. ix.
163. *Comptes rendus*, **143** (1906), p. 997.
164. Poincaré, *Science and Hypothesis*, pp. 178–179.
165. Ibid., p. 179.
166. Poincaré, 'The principles of mathematical physics', trans. George B. Halsted, pp. 604–622 in *Congress of Arts and Science. Universal Exposition, St. Louis, 1904*, ed. Howard J. Rogers (New York: Houghton, Mifflin, and Co., 1905), p. 610.
167. Ibid., pp. 621–622.
168. Arthur Hannequin, *Essai critique sur l'hypothèse des Atomes*, 2nd ed. (Paris: Félix Alcan, 1899). Hannequin was a *professeur à la Faculté des Lettres* at the Université de Lyon.
169. Émile Picard, *La Science Moderne et son état actuel* (Paris: Flammarion, 1905), pp. 117–122.
170. Abel Rey, *L'Énergétique et le mécanisme* (Paris: Félix Alcan, 1907).
171. Ibid., p. 80.
172. William Crookes, 'Les théories modernes sur la matière', *Revue scientifique* (4), **20** (1903), p. 225.
173. Review of Ernst Mach's *Mechanics*, in *Revue scientifique*, **20** (1903), pp. 787–788.
174. Henry Le Chatelier, 'Les phénomènes de combustion et la production de la puissance mécanique, de la chaleur et de l'électricité', *Revue scientifique*, (4), **9** (1898), p. 226.
175. Ibid.
176. In Ibid., p. 227, Le Chatelier wrote that 'We reject the expression "energie", despite its very common usage today in scientific language, because—like the expression "force"—it lends itself to misunderstandings . . . so-called free or utilisable energy is identical to motive power [the "puissance motrice" of Carnot].'
177. Prefatory remarks by the editor to Cornu, 'Quelques Mots de Reponse à 'La déroute de l'atomisme contemporain', *Revue générale des sciences*, **6** (1895), p. 1030.
178. Prior to Brillouin, Joseph Boussinesq had espoused the mechanical viewpoint in his lectures on mathematical physics at the Sorbonne, suffering heavy criticism

from Saint-Claire Deville and Berthelot. See, e.g., his 'Recherches sur les principes de la méchanique, sur la constitution moléculaire des corps et sur une novelle théorie des gaz parfaits', *Mémoires de l'académie des Sciences et Lettres de Montpellier*, 8 juillet 1872, esp. pp. 1, 18. In his article, 'L'évolution de la physique', *Revue scientifique*, (5), **5** (1906), pp. 483–484, Lucien Poincaré also discusses the adherence to the mechanical-kinetic picture by Verdet, Jamin, and Violle.

179. Marcel Brillouin, 'Pour la Matière', *Revue générale des sciences*, **6** (1895), p. 1033.
180. Ibid., p. 1034.
181. Wilhelm Ostwald, 'Lettre sur L'Énergétique', *Revue générale des sciences*, **6** (1895), p. 1070.

Chapter Two

Young Perrin: Initial Investigations

A number of years after the virulent controversy over the reality of atoms and molecules had subsided into a more amicable discussion about the complexities of sub-atomic structure, a still somewhat embittered Max Planck wrote that he himself had experienced during the 1880s and 1890s the inefficacy of a young scientist's attempts to win acceptance for a superior idea strongly opposed by powerful and older figures in the scientific community.

> An important scientific innovation rarely makes its way by gradually winning over and converting its opponents; it rarely happens that Saul becomes Paul. What does happen is that its opponents gradually die out and that the growing generation is familiarised with the idea from the beginning.[1]

One may indeed apply this analysis to the success of the molecular hypothesis and to its most dedicated experimentalist and propagandist, Jean Perrin. Of a different generation than Mach and Ostwald, Berthelot and Duhem, he initiated his education and research just as the concentrated work in cathode rays and X-rays, black-body radiation and quantum theory was beginning, and he quickly established a reputation and the professional ties that earned him audiences for whatever further work he might do. Supported by his early work on cathode rays and X-rays, thermodynamics and ionisation of colloids, he refuted the arguments of the older generation with new experimental data on Brownian movement, and he inculcated the younger generation with the empirically verified concept of molecular reality.

Before considering in depth the work that secured Perrin's scientific prestige and that consistently argued the physical reality of molecules, we must first seek to understand the man himself: who he was, what motivated him, what—outside of science itself—occupied his time and determined the extent of his scientific output, as well as its direction. As with many of the men he opposed, notably Ostwald and Berthelot, we discover a breadth of interest, a confidence in human reason, and an overpowering belief in an education that revamps both science and society. With Perrin, the 'heavenly city'[2] was a very real possibility, through a combined advance of scientific and spiritual evolution. In speaking of Perrin, the French physicist Louis de Broglie reminisced that,

> The true scientist . . . is always a believer, a man who believes in the value of science and the power of man's spirit to resolve the problems that observation of the external world and the constant increase of our knowledge ceaselessly pose. To that faith he joins . . . the hope that scientific progress . . . will be accompanied by a progressive appreciation of spiritual values, bound to a spiritual evolution reflecting the very tendencies of life itself.[3]

Biographical and Scientific Profile

Of modest family origins, Jean Perrin was born at Lille on the 30th September, 1870. His paternal grandparents had been rather poor peasants of Saint-Dié, and his father, an army officer who had come up through the ranks, was fatally wounded at the battle of Saint-Privat, near Metz. His mother, who was of a *petite bourgeoisie* family at Boulogne-sur-Mer (and who was related to Frédéric Sauvage, inventor of a first form of the marine helix) raised him and his two sisters.[4]

Young Perrin embarked upon his secondary studies at the *lycée* of Lyon and then entered a class of special mathematics at the Lycée Janson-de-Sailly at Paris, where Émile Lacour encouraged him to prepare for the École normale supérieure. He was accepted in 1890, but did not enter until 1891, after serving a year in the army. It was at the École normale that Perrin began his studies under Marcel Brillouin, and after the completion of his work he was offered a teaching position in secondary education. Refusing this, he was named *agrégé-préparateur* at the École normale, and he began his scientific research in 1895, supplied with a small scholarship procured by Jules Tannery.[5] Certainly, the financial difficulties of scientific study during his youth were in part responsible for his subsequent concern with governmental support for science in France.

In 1897, he received his doctorate with a thesis on cathode rays and X-rays and was subsequently made *chargé de cours de chimie physique* at

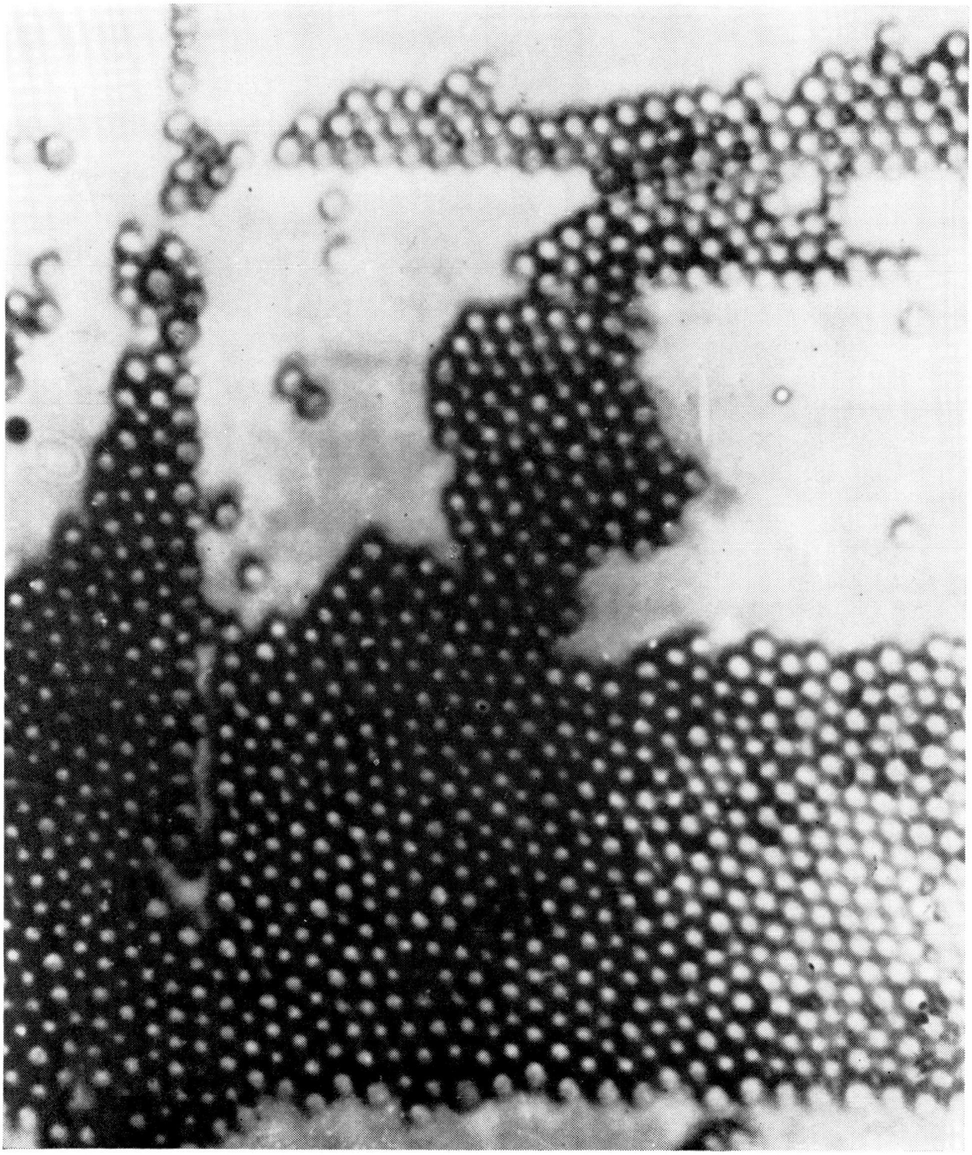

I. *A photograph of granules of gamboge utilised by Jean Perrin in determining their radius.*

II. *The Solvay Council of Physics 1911*. From left to right: *H. A. Lorentz*, O. *Warburg, Jean Perrin, W. Wien, Marie Curie, and Henri Poincaré.*

the Sorbonne. So successful was he in the organisation and teaching of this new course that a chair of physical chemistry was created for him in 1910, which he held until his death in 1942. In the next years, he was a member of the Solvay Conference (1911 and 1921), an exchange professor at Columbia University (1913), and an officer in the Engineers and adviser to the French government during the First World War. He was elected to the Académie des Sciences in 1923 and received the Nobel prize for physics in 1926.[6]

Shortly after completing his doctoral dissertation, Perrin married Henriette Duportal, the sister of a comrade and a baccalaureat graduate —the latter an unusual accomplishment for a woman at this time.[7] They had two children: Aline, who married the artist Charles Lapicque, and Francis, who followed closely in the footsteps of his father, receiving the chair of atomic and molecular physics at the Collège de France in 1946. Francis Perrin also did some work with his father on fluorescence in the early 1920's and was interested in mathematical studies of Brownian movement of rotation.[8]

Jean Perrin's scientific career falls roughly into three major periods of research, two of these before the First World War. The first involved research on cathode rays and X-rays and resulted in his doctoral thesis of 1897. Subsequently, while organising his course of physical chemistry at the Sorbonne, he dealt with thermodynamics, writing the textbook *Les Principes* (1901), as well as papers on osmosis, crystallisation, and ion transport. This led to an interest in the electrical charge on surfaces, and by 1905 he was studying the electrical properties of colloids—an investigation which flowered into his 1908–1911 papers on Brownian movement in his second period of major research. After some work with thin films, he published *Les Atomes* (1913), a book which culminated approximately twenty years' work, nearly all of which had delved into the discontinuous structure of matter.

His investigations interrupted by the war, Perrin served as an *officier du Génie*, and, like so many of his scientific associates throughout Europe worked in acoustics, designing apparatus to detect underground mines, planes and submarines. His basic research in this area was incorporated into an article published much later in 1932.[9] After the war, his third period of research commenced, with studies of fluorescence, the role of light in chemical reactions, and mechanisms of activation and de-activation of molecules. In collaboration with his son Francis, he affirmed that fluorescence is due to the return of particles from an excited state to their normal condition. Jean Perrin's motivation for initiating these investigations, however, had still been his concern with further proofs for molecular reality: 'I asked myself if some phenomenon wouldn't permit making visible, if not the molecules individually, at least fluctuations of concentration.'[10] Observing fluorescent solutions in very thin layers in special tubes and in thin films of soap bubbles, he failed to discover the fluctuations, but became interested in other problems.

In these years, too, Perrin closely followed other researches in the infant science of nuclear physics, and in 1919 he suggested that the condensation of hydrogen atoms into heavier atoms could account for the heat of the sun, and that, on the basis of mass spectrometry, the fusion of four hydrogen atoms could produce an atom of helium and a loss of approximately one per cent mass.[11] Such a source of energy, wrote Perrin in 1921, constituted a 'colossally larger reserve than the energy of gravitation to which Helmholtz and Kelvin have thought they could attribute the origin of solar heat'.[12]

Further, in the discussion following Ernest Rutherford's paper on the structure of the atom at the Solvay Conference of 1921,[13] Perrin suggested that the nucleus was more complex in structure than usually pictured and that, in fact, projectiles colliding with the nucleus actually participated in an intranuclear rearrangement.

> These very experiments of Monsieur Rutherford appear to prove that it is necessary to renounce this idea of a simple collision. The projectile—due to its great velocity and despite a very strong electric repulsion—can arrive, slowed down quite a bit, in the immediate neighbourhood of the nucleus. At that moment, a 'transmutation' is produced, probably consisting in an intranuclear rearrangement, with the possible capture of the incident nucleus . . . emission of the hydrogen nucleus forming the observed H ray, and perhaps with even other less important emissions.[14]

Perrin, also in 1921, proposed again his ideas on hydrogen fusion, which he had first broached in 1918, pointing out that calculations for a transmutation into helium would give 90 million years of solar radiation, whereas the older theories of the sun's energy allowed only 30 million. It was this discussion which led Rutherford to muse aloud,

> One disposes of quantities of heat far larger than according to the ideas of Kelvin and Helmholtz. . . . The idea has come to me that the hydrogen

> of nebulae provides particles which one may call 'neutrons' and which are formed of a positive nucleus with an electron at a small distance. . . . It is difficult to understand otherwise how some units of positive charge can penetrate into the nucleus, on account of the forces of repulsion.[15]

In the next years, Perrin's research activities slowed down, though he published several original papers, as well as books and treatises, each showing that he was closely attending the ceaseless advance of discoveries in physics, as well as their philosophical implications. Perrin's own time, however, was increasingly devoted to political and organisational duties. Such a combination of science and politics was not novel in France; Marcelin Berthelot, whose condemnations of atomism set the stage in France for Perrin's vindication of molecular reality, had himself served as senator-for-life from 1881[16] and as foreign minister under Léon Bourgeois in 1895.[17] Both Berthelot and Perrin, though opposed in their approach to scientific method, shared an overpowering optimism about the ultimate potential of science, as well as a dynamic solicitude for French education and an active belief that the scientist's debt to society extends beyond his professional accomplishments. In their specific orientation within politics, as within science, they differed in fundamental respects. Although Berthelot spoke of science as a base of republican politics and as the source of free thought, he remained silent on that index of radicalism within the Third Republic—the Dreyfus Affair.[18] Perrin, on the other hand, signed the first list of signatures collected in the *Quartier Latin* of Paris to support Zola's *J'Accuse*,[19] and with time, he became a thorough-going socialist, imbued with the conviction that a combination of scientific progress and socialist victory would ultimately liberate and transform humanity.[20] Wrote Léon Blum, prime minister under the Popular Front,

> Jean Perrin was a socialist, and no socialist, no scientist has ever been more attentive than Jean Perrin to the social significance of Science. . . . [He] wanted to put society in the service of Science. . . . He wanted, in return, to put Science in the service of social justice.[21]

Though Perrin's optimism about the contributions of science to society was extremely high, he never boasted that science alone, through its independent and automatic development, could endow the world with abundance, justice, and happiness.[22] He believed, rather, in the human being behind science, not in a mechanised juggernaut resolving mankind's problems by wit of a self-gathered momentum.

If the goals of science were to be the conquest of man's material environment and the liberation of man from ignorance and superstition, these goals were only secondary to that of '. . . allowing him a full blooming of all that which can make his life one of nobility and beauty'.[23]

An eager reader of the science fiction of H. G. Wells and J. H. Rosny-aîné, Perrin placed relentless faith in man's ability to become wise and good. He foresaw a social and cultural evolution of man, an evolution not attributable to chance, but to the very elemental 'volonté de vivre' of Lamarck and an evolution which man himself would work out.[24] His overriding confidence in man's capacity to be reasonable was expressed in a speech at Strasbourg on 14th March, 1937, before a popular *manifestation*. Some men, Perrin lamented, might call him a 'dreamer' for predicting the twenty-four-hour week for working men, but:

> I know very well that you will not pass these days in sleeping nor, as some people have abusively said . . . in going to the cabaret, or even in simply engaging yourselves happily in games or sports. Rather . . . some of you will carve furnishings or decorate his house; another, taking up horticulture, will cultivate his garden. And each of you—each will develop his soul and mind in the scientific comprehension of the Universe, in the cult of the arts and of beauty.[25]

When the Popular Front ascended to power in 1936, Perrin persuaded Léon Blum to recognise formally the importance of research and discovery in a country's development and to implement the simultaneous progress of science and society by the establishment of a cabinet position for science. Blum responded by the creation of the office of under-secretary of state for scientific research, and, with Perrin's concurrence, he appointed Irène Joliot-Curie to the post, thereby representing both French women and French science.[26] When Mme Joliot-Curie found the cabinet duties overly demanding of her time in the laboratory, Jean Perrin was asked to succeed her.

In proselytising Blum to install scientists in political power, Perrin was not at all espousing a government by scientists, as had the Saint-Simonians. Rather, he believed that scientists should place at the disposal of their government the scientific imagination, its methodological criticism and experimental controls, and its rational approach to problems and solutions. If Perrin was determined that the nation utilise scientific resources, he was, in turn, resolved that science have access to national resources—and with the greatest possible efficiency.[27] When he attended the cabinet councils at the Hôtel Matignon, he carefully followed the debates, though he rarely intervened.[28] Where he did initiate action through the cabinet and the assembly was in his variety of programmes for centralised governmental support of scientific research and of scientific education, and in this he was indefatigable.

Perrin had personally experienced in his laboratory at the Sorbonne the poverty of funds available for basic scientific research, especially that research which was not technologically oriented, and he realised full well the necessity for a central organisation. In 1933, he published a 25-page brochure, 'La recherche scientifique', and followed it in 1938 with a detailed plan for an agency in 'L'Organisation de la recherche scientifique en France'. He also organised legislative lobbies and used his own influence and that of his friends to attain, step-by-step through the 1930s, his goal. In 1936 the *Service nationale de la recherche* commenced full operation with a hierarchy of research scholars and directors of research, utilising both independent men and members of the higher-education establishment; in 1939 the completely centralised *Centre national de la recherche scientifique* emerged with sections for pure and applied science.[29]

Another of Perrin's contributions to the organisation of science in France was his persuasion of the Baron Edmond de Rothschild, whose friendship with Claude Bernard had inspired in him a great interest in the physico-chemical mechanisms of the body, to found, in 1927, the Institut de Biologie Physico-Chimique. Perrin and his friends Georges Urbain, André Mayer, and Pierre Girard were its first directors.[30] Finally, concerned, too, with staffing the organisations for scientific research which were being assembled, Perrin conceived one of his most original ideas, the Palais de la Découverte. In articles in 1937 and 1938 he pleaded for a public institution to educate and stimulate the youth of France into scientific knowledge and discovery:

> In the Palais, we want to realise a vivid exposition where, as much as possible, the fundamental discoveries which have enlarged our intelligence will be repeated in spectacular fashion with the most modern resources.... The experiments will be re-enacted under the eyes of visitors by 'demonstrators' who will explain them with the aid of synchronised records or cinematographic films. Or on the other hand, brief commentaries, in the form of bulletin boards, will relate the experiments.... Thus the general public ... will understand that ...

> if one wants to draw great results, then this is precisely when discovery should be pursued without practical preoccupation.[31]

Perrin hoped, too, that young people would discover in this museum of historical laboratories a new calling, as Faraday had done when conducted by chance into the Royal Institution.[32]

In June 1940, when the Germans invaded France, Perrin was at Bordeaux with the C.N.R.S. Shortly before the Armistice, he departed for North Africa with those who hoped to organise a government of the French Resistance.[33] When this failed, he returned to Lyon, where he was advised to leave France, as his membership in the *Comité de vigilance antifasciste* and his well-known opposition to the Germans might well endanger his life.[34] It was also urged that he might do more for the French cause if he were free to speak in the United States, where his son Francis was teaching at Columbia University.

Once in New York, Perrin established, with Henri Focillon, the French University of New York and attempted to clarify American opinion on the sentiments of the French people regarding the men of Vichy.[35] He never lost faith in the ability of the French to defeat the Germans, nor did he falter in his fundamental optimism about the ultimate future of man. The Germans he regarded as an aberration of humanity to be conquered by men of high ideals. Sadly, he did not live to see his faith vindicated, but died in New York on 17th April, 1942, at the age of seventy-two.

The Education of Jean Perrin

In devoting his life to the progress of science and society, Perrin always seemed to concentrate upon certain issues which he judged central to that progress. Louis de Broglie termed Perrin a 'strategist', one who surveys a situation at a glance and then designs a crucial experiment to decide the basic question.[36] Certainly, in regard to the mutual service of science and society, Perrin had deemed the creation of the C.N.R.S. a critical need. Within science itself, he had begun to feel early in his youth that a decisive step lay in the proof of the physical reality of molecules and atoms. In fact, he first saw this problem clearly delineated after he undertook his studies at the École normale in 1891, and before the turn of the century he had stepped forward to take a strong hand in the philosophical and scientific struggle then raging.

The period of Perrin's education at the École and of his first years at the Sorbonne in the 1890s was that of the intellectual turmoil of the *fin-de-siècle*, as discussed earlier. In the political arena, the Dreyfus Affair was to grip a France already crippled by severe economic problems and protests for socialist reform. The anti-positivist, anti-scientistic condemnations of Brunetière and his followers were flowering full, and a deep pessimism gripped intellectual France. The attempts to mould a new, clearly defined science, free from hypothesis and from the mechanical models of the past, were being butted about throughout Europe, while in England the experiments of Crookes and Thomson on cathode rays posed new problems of interpretation.

When young Perrin first arrived in Paris to attend the Lycée Janson-de-Sailly, he passed many hours with Armand Duportal, whose grandfather had been a political prisoner and whose engineer father was an ardent republican.[37] In the next years, Perrin joined the small circle of intellectuals who met at the Librairie Bellais-Peguy with Lucien Herr and Charles Andler—all converts to socialism. While Perrin worked in the laboratory and lectured at the Sorbonne in the nineties, he was ever atune to 'le sourde rumeur' in the streets.[38] When Jean Jaurès called for a new *Encyclopédie* for the proletariat, chronicling the movement of human thought from Kant to Renan, from Hegel to Marx, and '. . . the movement of science, that went from Laplace to Maxwell, passing by Darwin',[39] the young physical chemist formed (or found further support for) ideas that he maintained the rest of his life. Embracing Jaurès' social thought with enthusiasm, Perrin aligned himself among the groups of the left without adhering to any particular political party.[40] As mentioned earlier, when signatures were sought in 1898 to protest the government's manipulation of the Dreyfus Affair, his name was on the first list.

When Perrin met at the *Librairie* with his friends, questions of government and socialism were not the only subjects posed, however. Perrin's *confrères* Paul Langevin and Noel Bernard were among the group, and the conversation certainly shifted to their researches and to current questions in the physical and biological sciences. Among the other companions of Perrin in these years and in those to follow were many leaders in French science, including Georges Urbain, the discoverer of the rare earths; André Job, the investigator of mechanisms of chemical reaction; and the mathematicians Paul Painlevé and Émile Borel. Also in his intimate circle were Marie and Pierre Curie (and later the Joliot-Curies) and André Debierne, who first separated actinium in 1899 and then isolated radium with Marie Curie in 1910. The Perrin home neighboured that of the Curies, and when various combinations of the friends gathered at meetings or Sunday dinners, talk was of science, the Dreyfus process, the dramas of Wagner, poetry, and social doctrines.[41]

By 1900, one could be certain that the researches of his friends greatly influenced Perrin's concept of molecular reality and had intensified his conviction in its truth.[42] These researches included the investigations by the Curies and Debierne of radioactivity, the work of Langevin on the ionisation of gases, and the preoccupation of Job with unstable intermediates. Like molecules, such intermediates were highly hypothetical, and their real existence had not yet been experimentally demonstrated.[43] Perrin's peers, however, were not the most responsible for his conviction of the critical nature to science of the proof of molecular reality. This role, rather, was played by his educational milieu and his teacher at the École normale, Marcel Brillouin.

Perrin arrived at the École normale shortly after the period when Maurice Caullery had reported a sharp scepticism about atomistic notions among its teachers. Henri Saint-Claire Deville had died in 1881, and in the following decade it became more and more admissible to conjecture about atoms. Berthelot, of course, was still quite active and maintained his chair at the Collège de France, while the other prominent living French chemist opposed to atomism, Henry Le Chatelier, carried out experiments at the École des Mines. Duhem was at Bordeaux, physically remote from Paris, though publishing in influential journals.

As discussed earlier, the controversy over atomic reality was part of a larger discussion of positivism and idealism, scientific method and the value of science occurring throughout much of Europe, but most viciously in Germany and France. The debate which least swayed the pattern of Perrin's intellectual development was probably the polemic concerning the so-called 'bankruptcy of science', not because he was unaware of it or even because he thought it unimportant, but because it reeked of a certain extremism which was anathema to his mentality.

If one will recall, Paul Bourget raised the question of the morality of scientific doctrines, such as determinism and fatalism, in his novel *Le Disciple* (1889), and, in 1895, Brunetière—who was a *maître de conférences* at the École normale—brought many a reader to his feet with an indictment charging science with broken promises and implicit immorality. It may well have been that the polemical articles of Brunetière, Richet, Berthelot and others provoked as much controversy in France of the 1890s as has the recent quarrel over the 'two cultures'.[44] The British journalist, Wickham Steed, then a close friend to Charles Andler, Lucien Herr, and Charles Seignobos—all associates of Perrin—reported that Brunetière's conversion from Taine to the church '. . . was greeted with indignation by the Latin Quarter where the students prevented Brunetière from giving the lecture he was to deliver before an elegant audience in the Sorbonne'.[45]

As individuals aligned behind Brunetière or Berthelot, Perrin, in truth, could not himself unreservedly support either man. Certainly he had little sympathy with Brunetière, since his own faith was not in the idealism of the church, but in the materialism of socialism, which, in turn, placed strong faith in man, rather than derogating him, like religion. Perrin vehemently assessed the 'narrow and savage obscurantism' of supernaturalism a number of years later, recalling:

> . . . the puerile cosmogonies which all religions have imposed, dating from the times when the earth seemed the centre and almost the whole of a Universe organised for men. . . . Then, and not only among savage people creating singular taboos, but in the very breast of our European civilisation, intolerance and superstition enslaved all *esprits* under terror, threw men against each other, justifying, sometimes sanctifying massacre and torture, in order to impose uncertain ideas, arbitrary rites and incomprehensible dogmas for which the word belief could not, in fact, have any meaning.[46]

In fact, young Perrin, in analysing proofs for the existence of God, found that the word 'God' had no significance for him. Without really suffering a grave crisis, he lost the religious beliefs with which he had been raised, '. . . in discovering that a thought free of all dogmatism and a morality founded on esthetics and kindness are without doubt the most elevated expression of the human conscience'.[47]

The Comtian positivist position expressed by Berthelot was also unacceptable. Perrin later recalled that his friend Pierre Curie remarked, in discussing 'the social question', that scientists tend to suppress the problem of poverty and inequality in the belief that they will create enough riches for everyone. 'Yet', Perrin insisted, 'let us recognise that the increase of riches imposes a social reorganisation which is resoluble certainly, but difficult. . . . We are well-informed, moreover, and we are able to act: this will be the great task of the sociologues and the political men, that of being consecrated to these great problems and avoiding the dramatic fumblings of undirected rearrangements'.[48] Such was his agreement with both Jaurès and Blum, who criticised Berthelot's faction for believing that science alone could revolutionise the social order without militant human action. 'Yes . . . it [science] creates the possibility of new social forms. But it creates only the possibility.'[49]

Accompanying the reaction against the Comtist conception of science—a reaction which questioned the morality and social responsibility of science—was the philosophical, but essentially intrascientific controversy over methods within the scientific discipline. The central questions of the logical positivists focused on the role of intuition in scientific theory, the relative merits of the inductive versus the deductive method, and a definition of the final form of theory or of a body of theories. Perrin's first book, *Les Principes*, published in 1903, demonstrated that he was fully cognisant of the trenchant arguments in these debates. As he himself preferred experimental to theoretical science, publishing exclusively experimental researches with limited mathematics, his viewpoint was necessarily moulded by this predisposition. In later years, he said that '. . . in reality, there are no methods particularly characteristic to physical chemistry. There is a physico-chemical *esprit*, an attitude of the experimenter managing matter following the procedures of the physicist or chemist.'[50]

As an experimentalist, Perrin's attitude seems to have corresponded with that of Claude Bernard:

> Feeling gives rise to the experimental idea or hypothesis. . . . The whole experimental enterprise comes from the idea, for this it is which induces experiment. Reason or reasoning serves only to deduce the consequences of this idea and to submit them to experiment. . . . There can be no method for making discoveries, because philosophic theories no more give inventive spirit and aptness of mind to men, who do not possess them, than knowledge of the laws of acoustics or optics can give a correct ear or good sight to men deprived of them by nature.[51]

The full complex of Perrin's reactions to the discussion over scientific method can better be seen in the more complete discussion of *Les Principes*, given below.

Another of the controversies which enveloped science as Perrin was studying at the École normale was, of course, the general debate involving the validity of the kinetic theory, the appropriateness of mechanical models for describing the causal reality behind appearances, and the actual existence of discontinuity in nature. Do atoms exist? If so, is their reality capable of proof, and if not, can one justify their employment as a hypothetical device? While debates on these issues warmed up, Perrin himself entered the École normale and became an admiring pupil of the man who publicly denounced Ostawld's 'déroute' of atomism in 1895 and who was to write the preface for the French translation of Boltzmann's lectures on gas theory.

Marcel Brillouin had become *maître de conférences* at the École normale in 1887, where he remained until obtaining a chair in general and mathematical physics at the Collège de France in 1900. He wrote two doctoral *thèses*, one in mathematics (1880) and the second in physics (1882). His first important mathematical treatises were in hydrodynamics,[52] and in the late 1890's he was publishing papers on gaseous diffusion from a molecular standpoint. In his *Recherches récentes sur diverses questions d'hydro-dynamique*, published in 1891,[53] he wrote regarding the value of hypothesis:

> It is always easy to make an ordinary hypothesis plausible in its general features: but this is only an exercise of the imagination in scientific utility. It is necessary to squeeze out the consequences, to attain numerical laws, to compare them to experience. This work, should it destroy the hypothesis, will not have been useless; it enriches the spirit of new views on a whole order of questions. . . . A hypothesis is only a means of forcing thought to renounce its habits; for being useful, it is necessary that it should be original and susceptible of precision.

One can be sure that in his years of tutelage under Brillouin, Perrin became quite familiar with the work of Maxwell and Boltzmann, as well as with the equations and conditions of thermodynamics. Brillouin himself was quite absorbed in the mechanical models of Kelvin, especially the 'atomes-tourbillons', and with the ether-and-matter theories of Lorentz and Larmor.

Fully aware of the scepticism about the validity of kinetic theory and the criticisms of the usefulness of the molecular conception, Brillouin persevered in his belief that the corpuscular hypothesis promised greatest fecundity in solving problems of physics and chemistry. To those who criticised this hypothesis for its simplicity, he answered that the complexity of the problems posed to the scientist is in the nature of things, and that a fundamentally simple idea may very well serve to group the large number of observed phenomena, if—as with the concept of molecularity—the logical analysis of the hypothesis' contents leads to a richness of associations and combinations.[54] Brillouin esteemed Boltzmann's statistical treatment of gas theory a prodigious achievement, although he perceived many difficulties yet to be surmounted.

ON Boltzmann

> Certainly the molecular hypothesis is prodigiously simple and fertile on the whole, but it is none the less difficult to put into operation. The obligation of attaining some average results, some quantities observable by themselves, forces the use of methods of statistics and probabilities; but our ignorance of the precise properties of molecules and of the law of molecular action gives rise to doubts on the correctness of some suppositions made in the course of calculations.[55]

Nevertheless, exulted Brillouin, at the very moment when the energeticists were arguing (*dissidente et niaient*) the fertility of the molecular hypothesis and kinetic theory in physics, they had reappeared in chemistry in the electrical experiments of J. J. Thomson and his school at Cambridge.[56]

Cathode Rays and X-Rays

Considering the interests of Brillouin and the direction of the researches of his friends, it is not at all surprising that Perrin's first experimental work, undertaken in the laboratory of Brillouin and Jules Violle, was with cathode rays. In writing an *avant-propos* for his *Notice sur les Titres et Travaux* (1918), Perrin characterised his career in two sentences: 'My works have been especially of the experimental order and are attached to Atomistics. However, I have been led by my teaching [*enseignement*] to publish an exposé of Energetics.'[57] It was altogether typical, then, that his first researches dealt with a new experimental problem, the consequences and implications of which were of a profound and general nature for atomic and molecular theories. As his friend Pierre Girard later recollected, '. . . his taste of the Universal and . . . his profound intuition of the real' were two of Perrin's most consistent characteristics.[58]

In attacking a scientific problem, his attitude was a far cry from the strict economy of a Mach or the systematic classification of a Duhem, for he placed great confidence in the imaginative faculties—in particular, the atomic-molecular hypothesis—as did Boltzmann and Brillouin. The scientific researcher, Perrin summarised, has

> . . . some qualities of the technician: precision, manual dexterity, prudence mixed with audacity, and that imagination which understands profound analogies or knows how to devine the invisible. . . . But these qualities of *esprit* would be sterile without the motivating qualities which are like the soul of the *chercheur*, developed from a profound taste for Mystery to which risk even gives a charm, from passionate desire for Adventure, and from that love of Poetry and Beauty without which there is no artistic creator.[59]

For his part, Perrin's primary goal from the very beginning of his scientific career was to prove the reality of the invisible atom, to eliminate as a 'puerile anthropomorphism'[60] those strictures which seemed logically necessary to many others. His penchant for experimentation and his scientific interests were strongly enough mixed with a firm faith in molecular reality to be noted by his students. One student wrote of Perrin that, in his lecturing, '. . . in a familiar gesture, with his fingers joined, as for a prayer, with hand raised in the form of a fist, he looks like a naïve figure of the thirteenth century, descending from his altar, come to preach while in a trance a science of divine inspiration. He "sees" atoms—there is no doubt at all—as Saint Thomas saw seraphims.'[61]

The actual visibility of molecules seemed possible to many for the very first time in the experimental studies of cathode rays. For a number of years before Perrin entered the École normale, brightly coloured effects had been observed in highly evacuated tubes through which an electrical discharge was passed. Opinion over the nature of the cathode discharge fell into two segments, roughly divided on national lines. The English physicists, including William Crookes and J. J. Thomson, supposed that the rays consisted of negatively charged molecules (ions) in what Crookes termed a 'fourth state of matter,' while the Germans, including Goldstein, Hertz, and Lenard, were convinced that the cathode rays were some form of electromagnetic radiation, similar to light.[62]

To Perrin, the decisive point in the controversy seemed to be whether or not the cathode rays were, in fact, charged, since both the wave and the particle conceptions explained most of the properties of the cathode rays: their propagation in straight lines and production of shadows of opaque objects, their deflection by an external magnetic field, their fluorescence, and their ability to be focused or diverged by a concave or convex cathode. What particularly distressed Perrin were the experiments of Lenard showing that the cathode rays penetrated thin glass sheets. Perrin deemed it imperative to determine whether or not the cathode rays were charged. He reasoned that if they were charged particles, they might differ from ordinary matter in size and velocity, thereby passing through Lenard's windows. Fortunately, Perrin was not acquainted at this point with Hertz' 1883 experiments,[63] showing that the cathode rays were unaffected by electrostatic forces and were therefore uncharged.[64]

In 1895, then, Perrin's first published article appeared in the *Comptes rendus* of the Académie des Sciences,[65] describing his successful attempt to detect a charge on the cathode rays in a Crookes' apparatus (see Fig. 1 below).[66]

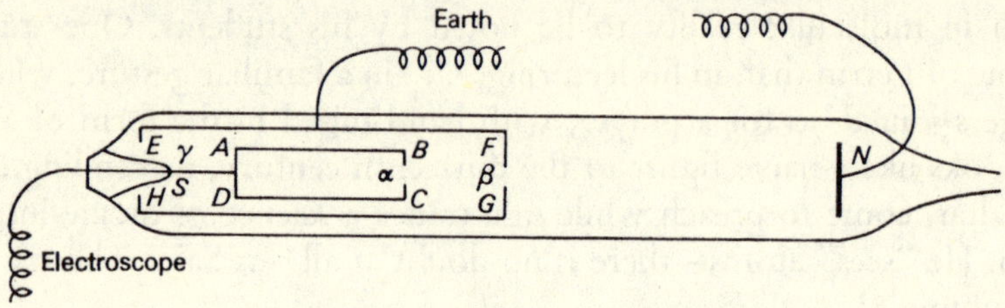

FIG. 1

In the experiment, Perrin wrote,

> The electrode *N* served as cathode; the anode was formed by the protective cylinder *EFGH*: a pencil of cathode rays then penetrated into the Faraday cylinder. Invariably, this cylinder was charged with negative electricity.[67]

When the Crookes tube was placed between the poles of an electromagnet, the cathode rays swerved away from the cylinder and a charge was no longer registered on the electroscope; once the electromagnet was deactivated, the charge again appeared. Concluded Perrin, '. . . *the cathode rays are charged, then, with negative electricity*'.

In this initial paper, Perrin also found what he termed a positive electricity when the cathode and anode were reversed. He later explained this as an error, for when the slit *B*, of several millimetres diameter, was replaced by several holes of small diameter—with total surface comparable to the surface of the first opening—the positive electrification in the Faraday cylinder failed to appear. He assumed that the mistake was due either to a strong drop of potential in the neighbourhood of the cathode or to other causes.[68] It was this 1895 paper which J. J. Thomson used in his 1897 paper on cathode rays, definitively establishing the synonymity of the current and the rays. To dispel any doubt that the current followed the exact path of the rays, he modified Perrin's apparatus to bend the rays around corners.

The 1895 paper was not the only work that young Perrin did on on cathode rays, however. In further experiments to establish the charge on the rays and the deviation of the cathode rays in an electric field, Perrin developed the method of *potentiel-retardateurs* which could be used for measuring the actual velocity of the cathode radiation. In the apparatus shown below (Fig. 2),[69] the rays issued from the cathode N and crossed the anode P, which was formed by a metallic cloth connected to the positive pole of an electric machine M. They then moved toward the auxiliary electrode N', which was covered by a fluorescent powder and was connected to the negative pole. When the difference of potential was zero, a vivid fluorescence was seen at N', but as soon as the machine was activated, the fluorescence gradually paled and then

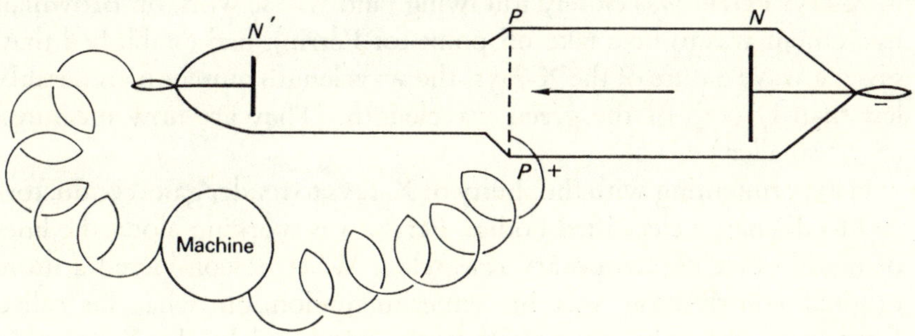

FIG. 2.

disappeared, reappearing as soon as one discharged the machine. Hence, the electric field $N'P$ opposed the movement of the cathode rays; Perrin also found that the strength necessary to stop the rays varied with pressure.

J. J. Thomson had already discussed the relation $eV = \frac{1}{2}mv^2$, and Perrin began to measure e/m, where V was the potential of the electric field parallel to the cathode rays and completely stopping their movement. 'But', he later wrote, 'as I began to make these measurements, I learned that J. J. Thomson (who, after having repeated my experiment, found himself equally led to measure e/m) had preceded me.'[70]

Throughout his researches, Perrin maintained that the rays were probably ions, and he noted that their velocity was much higher than that which kinetic theory assigns to ordinary molecules of gas. Since he assumed that the cathode-ray corpuscles must be quite small in order to penetrate Lenard's windows, he no doubt would have come to the same conclusion as Thomson regarding their sub-atomic nature, had he been the first to calculate the e/m ratio.

Along with these investigations of cathode rays, Perrin began researches on X-rays, which had been discovered by Röntgen late in 1895. When Perrin published his first paper on X-rays in January, 1896, little was known about them, other than that they pierced all but the denser materials, that they originated where cathode rays struck, that they were not deflected in a magnetic field (as were cathode rays), and that they discharged electrified bodies. Röntgen detected no refraction or interference phenomena in his first experiments. In Perrin's initial studies, he largely repeated Röntgen's work and postulated that if the phenomenon were periodic, the wavelength of the X-rays must be less than that of green light. By July, 1896, Léon Gouy, whose researches in X-rays Perrin was closely following (and whose work on Brownian movement was to be a take-off point for Perrin), had established that, given a wave nature of the X-rays, the wavelength must be considerably less than 1/100th of the green wavelength. (They are now measured at 10^4 smaller.)

In experimenting with the ability of X-rays to render gases conductors and to discharge electrified bodies, Perrin was working along the lines of many other contemporary researches. What he considered a more original contribution was his experimentation on what he called *l'effet métal*, whereby an ionisation was produced by the X-rays at a gas–metal interface. Perrin referred to the *ionisent* power of X-rays both in connection with gaseous and metallic effects, though he hastened to explain in a footnote that 'The word *ionisation* does not imply any molecular hypothesis, and should evoke only the idea of a separation of opposite electricities, linked to the nature of matter.'[71]

Yet, much like Newton in the appendix to the *Optics*, Perrin inserted a section into his experimental work which postulated what he had not proved. This section leaves little doubt, especially in the light of his future writings, of Perrin's own preference in the controversies between anti-atomism and the molecular hypothesis, and between energeticism and kinetic theory.

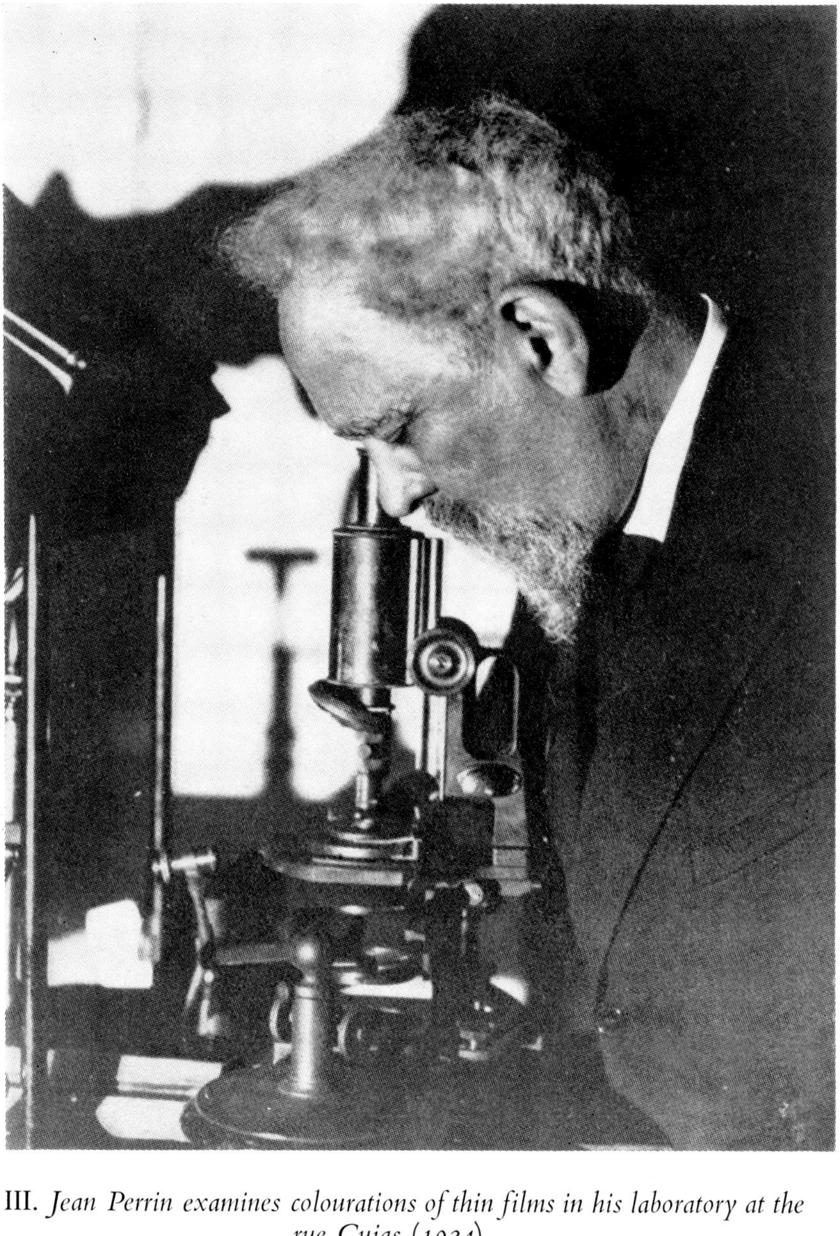

III. *Jean Perrin examines colourations of thin films in his laboratory at the rue Cujas (1924).*

IV. *The last photograph taken of Jean Perrin (New York, 1941).*

> Until the present, I have avoided any molecular hypothesis. But even if it is not necessary, it can be interesting to suppose that, upon their passage into the gas, the Röntgen rays break up certain molecules. The ions thus formed, charged with opposite electricities by the very fact of their separation, could then be moved in the gas . . . Since, in the kinetic theory of gases, the internal energy of a molecule is independent of the pressure and proportional to the absolute tempertaure, one states then . . . that, for each gas, the number of molecules broken up is proportional to the number of molecules encountered . . . and proportional to their mean energy.[72]

In making this postulate, Perrin publicly gives his earliest direct defence of the molecular hypothesis and becomes one of the first to observe that X-rays make gases conductors and discharge electrified bodies by creating ions or other charged bodies.[73]

A Reconsideration of Mechanics and Thermodynamics

It is difficult to say in what direction Perrin's researches would have continued, had he done nothing but experimental work in the next few years. However, when the Sorbonne authorised him to teach a new university course—that of physical chemistry—in 1898, he was compelled to slow down the pace of his research and to devote time and energy to the organisation of the course itself and of a laboratory of physical chemistry. He first visited laboratories in Germany, Holland, and England and then structured his course, writing the textbook *Traité de Chimie Physique. Les Principes*, which was to be the leading volume of a series of texts in physical chemistry (a series which he never found time to complete). The laboratory was in the rue Cujas, and, according to a pupil, was so high in the Sorbonne, that it was almost under the roofing. One entered the laboratory by a very narrow and dark corkscrew staircase, a veritable deathtrap (*caisse-cou*), which led to a landing for a string of four rooms. Continuing the ascent, the stair climbed to a vast attic (*grenier*) where the young students worked. Perrin himself occupied the largest of the four rooms below.[74]

At this time in France, scientists at the Sorbonne were unanimously discontent with what they considered the mediocrity, even the penury, of the means at their disposal, and the problem of balancing laboratory budgets was no small thing.[75] One of the most significant achievements of Perrin was his effecting the transferral of the rue Cujas laboratory to a new building with modern construction and equipment, which he, in large part, designed. This laboratory was completed in 1924.[76] It was in this laboratory that Monday-afternoon meetings, remembered by all of Perrin's students, were established. Here Perrin listened with sincerity and interest to the researches and problems of the younger workers, he himself arriving for the sessions with arms full of flowers from the *place de l'Odéon* while others prepared '. . . the tea, whose water boiled in a glass flask on a metallic trellis spreading out the flame of a Bunsen burner'.[77]

When Perrin's course was established in 1898, there was no chair of physical chemistry and he was only *chargé de cours*. So successful was he, however, that a chair was created in 1910, which he held until leaving France in 1940.[78] According to another of his students, Perrin's course contrasted strongly with the more traditional ones at the Sorbonne, in that there were no examinations.[79] By and large, the contents of the course were aimed at advanced students who had already obtained their first degree.[80] In the beginning only a few students were enrolled, with occasionally only one or two present, The number rose rapidly, and by 1905 the original amphitheatre was no longer adequate in size.[81]

Perrin's lectures lacked many of the characteristics generally attributed to a *maître de l'enseignement*. Beautiful order, clear presentation, perfectly logical construction were absent. Rather, students found in him other qualities most appealing and most challenging. One said that students discovered in his teaching the poetry, the stirring force, the philosophy of science vainly sought elsewhere.[82] Another recorded in homage that

> That which seduced me immediately with professor Jean Perrin is, I must say, his originality . . . his need of rethinking questions for himself, of rediscovering them in some way, of making them his own, of incorporating them and formulating them to his fashion. For that matter, it is not only in his verbal teaching, but in his books, that this need of possession and mastery by his own spirit are demonstrated. *Les Principes*, this is thermodynamics entirely re-thought by Jean Perrin.[83]

Indeed, amidst the barrage of academic and popularised textbooks of thermodynamics then being published, Perrin's *Les Principes* (1903) stood out as an original contribution. While Perrin always stressed the roles of intuition and hypothesis to his students, this was not an imagination without licence, but one strictured by the rules of logical reasoning. His treatment of thermodynamics grew out of his own education in a turmoil of controversies—an education broadened, no doubt, by metaphysical arguments, but also confused by them. His subsequent writings adhered to the analytical tradition taught by Marcel Brillouin and by Jules Tannery at the École normale—that tradition in which Tannery once remarked, 'Mathematicians are so used to their symbols and have so much fun playing with them, that it is sometimes necessary to take their toys away from them in order to oblige them to think.'[84]

Perrin reacted positively to this tradition and against the viewpoint of retaining physical theory, unmindful of its history. He rejected any scientific method whereby the steps to a theory or equation are discarded and only the final result retained. Perrin's attitude of ever being mindful of original derivations and definitions stylises his approach in *Les Principes* and may be viewed again and again, as in his preface to the 1929 edition of *Les Éléments de la Physique*. There he explained,

> . . . I have attempted to distinguish that which is only intellectual habit and that which is logical necessity, and I will be happy if some young reader, who will perhaps consequently realise new progress in this way, is struck with the idea that there is no notion which remains indefinitely valuable when one is separated more and more from experimental conditions where it is formed.[85]

In the preface to *Les Principes*, he cautioned that, '. . . in spite of all my enthusiasm for action, I think that it is better to understand before acting' and that to mask difficulties under the pretext of going faster is like putting blinders on a horse in order to get him to cross a ditch.[86] Defending his emphasis upon axioms, definitions, and logical deduction in *Les Principes*, Perrin insisted that, '. . . we are to understand the importance of methods which make judgment more correct and more clear, and we will agree that it is good to have taken up the habit of great precision in language and thought in relatively easy questions.'[87]

That logical reasoning had more than a scientific value and, in fact, was a large part of the educative value of a scientific culture was not lost upon Perrin. To have once recognised that certain phrases are in reality senseless or meaningless constitutes a step in the struggle against 'l'abdication de notre raison'.[88] *Les Principes*, in reality, is not merely a technical textbook written for the scientist, but, in part, a lesson in methodology, with applications for the cultural, social, moral, and political milieu. Perrin stresses the employment of logical necessity, rather than a slovenly default into intellectual or emotional habit; he insists that, aside from fundamental definitions which must be accepted, all must continually be questioned.

Perrin's preface to *Les Principes* is penned in defence of the inductive method, a defence which he feels necessary since the textbook employs primarily the methods of deduction. Citing the hypothesis of microbes in pathology—which was unverified for many years—he maintains that the value of hypothesis and induction has oft been proved. Like Boltzmann, Perrin insists upon the complementarity of the two scientific methods, pointing out that 'it is not only in the moral world that intolerance carries with it its chastisement'.[89]

Defending kinetic theory and atomistics against the energeticists' charges, Perrin points out that mechanical models are not unique in suffering from problems in semantics and logic; on the contrary, thermodynamics, whose principles he is to expound, suffers from many of the same weaknesses alleged against mechanical models. In discussing thermodynamics, one teaches in turn '. . . by an ironic *rapprochement* of difficulties of the very same order, both the elementary demonstrations of Mechanics and the reveries of metaphysiciens'.[90] Among the weaknesses which may be discerned in both contemporary mechanics and thermodynamics is, for example, the rigid insistence upon absolutes. The rule that heat cannot pass from a cold body to a hot one, 'as if one is speaking of an apple which passes from hand to hand' is almost metaphysical in its rigidity. Another example of metaphysics in thermodynamics is the inextricable confusion in laws involving the words 'force' and 'energy,' resulting in a science that is almost religious teaching by an energy cult. In contradistinction, wrote Perrin, 'I have sought what physical reality serves to support [energetics], without being content with the superfluously vague affirmation that "something remains constant".'[91]

Perhaps the greatest difficulty in thermodynamics was the very fact that it was a new discipline combining two older ones. Traditionally, as Perrin viewed it, the notion of discontinuity had been as fundamental in chemistry, as the notion of continuity in physics. In chemistry, it had been found that pure bodies could not be ranged by insensible gradations into a continuous series, and the law of definite proportions and of proportional numbers, for example, made sense only because of this discontinuity. In addition, the very criterion of a chemical phenomenon was that it be not only discontinuous, but also irreversible. On the contrary, the notions of continuity and reversibility were fundamental in physics. Along with continuous functions were imposed notions of equilibrium and reversibility, for instance, with the change of a gas to a liquid at the critical point. Physicists had been unconcerned with the nature of the body when viewing a phenomenon, and a physicist of the first calibre might know nothing of chemistry.[92] Physical chemistry, however, is a science in which 'all research and all theory, in order to be conducted or discussed, exist such that the same mind recognises in them, or at least in their major lines, both pure Physics and pure Chemistry'.[93] As in analysing the deductive and inductive methods, Perrin seems here to be stating that a meeting of two viewpoints is imperative for the survival of all. Again and again in his work the dictum appears that one view is not necessarily any more appropriate than another, that they may be two aspects of the same thing.

In his exposition of the principles of physical chemistry, Perrin deals with the large number of concepts which would be expected: there are chapters on force (statics and dynamics), on what he calls (in Ostwaldian terminology) 'factors of action' (temperature, pressure, volume, etc.), on the characteristics of stable equilibrium, the identification of pure bodies and their laws of combination, and the phase rule, in addition to discussions of the first and second laws of thermodynamics. The latter part of the book makes ample use of Gibbs and includes the contributions of the anti-atomists to equilibrium theory in the form of Le Chatelier's principle and Duhem's thermodynamic potential function. Berthelot's law of maximum work is also treated and criticised as imprecise except at low temperatures.[94]

In the first part of *Les Principes*, before proceeding to the theories of physical chemistry *per se*, Perrin discusses statics and dynamics in general. He proceeds with greatest care, ever aware of the vast array of criticisms and disputes regarding the logic of Newtonian mechanics. Foremost in his mind are the critical positivists' attempts to reduce mechanics to a simple description of phenomena and the demands by Kirchhoff, Hertz, *et al.*, to eliminate what they regarded as the mystical concept of force—a concept which, according to Duhem, Hertz had only succeeded in replacing by hidden masses and hidden movements. Hence, in introducing the principles of mechanics to the reader, Perrin adopts the logic-oriented, definition-conscious approach of the earlier critics. Yet his basic scientific attitude follows the line of Boltzmann in general, making use, for example, of Boltzmann's formulation that

> The conceptual symbols which we form have existence only in ourselves, and we do not know how to measure exterior appearances, but by the measure of our own representations. We are thus able to formally reject questions such as these: Does matter exist alone and is force only a property, or does force exist independently from matter, or—on the contrary—is matter a production of force? These questions do not have any significance, for all these concepts are only images of thought having for their object the correct representation of phenomena.[95]

For example, in discussing the vector used in an equation to signify the weight of a body, Perrin adds, 'It is not useless to add that there is no interest in deciding if the vector thus introduced represents a real or fictional force: here it is a simple question of language.'[96] Similarly, he notes in a treatment of magnetic forces that the magnetic mass is not a real entity, but only a calculating device; and, again, he remarks that while 'centrifugal force' is customarily termed a fictitious force and 'weight' a real one, this is '. . . a convention of language which does not appear justified to me'.[97] That historical usage of a term could be retained even when modern experiment indicates an anachronism in that usage is apparent in statements about the 'quantité de chaleur cedée': 'Here one has expressions which come to us from the time when the *materiality of caloric* was believed in. No one doubts that in the current state of things these expressions are incorrect; however, we employ them since they are imposed by usage.'[98]

In characterising Newton's law of universal attraction, Perrin words his statements most carefully, specifying that the equation indicates that bodies act *as if* an invisible filament with tension ($K.mm'/d^2$) were extended between two masses: that all occurs *as if* they were attracted to each other.[99] His manner of defining force is typical of the care which he lavishes upon all fundamental definitions and axioms, as he first considers a wire (*fil*), its tension and the 'force' acting upon a material point where the wire is attached; then, the force of a liquid on a solid body in terms of filaments attached to the body at a large number of points. The weight of a body is demonstrated by a vertical filament attached at a point of a carapace enclosing the body—a carapace composed of 'forces of contact', which maintain the body in undeformable shape. In considering force, Perrin insists, one must advance slowly in order gradually to enlarge the significance of the term, '. . . . in seeking not to jump the intermediaries; in order to always avoid force being able to appear as a sort of mysterious entity.'[100]

In building up an edifice of mechanics and thermodynamics, Perrin follows the traditional path, far afield from the attempts of Ostwald and Helm to reconstruct mechanics from thermodynamics. Perrin also rejects any dogmatic insistence upon absolutes, whether absolute movement or the absolute direction of that movement; he insists, as Boltzmann, upon the fallibility of scientific theory and the indispensability of revamping or augmenting it, but never of simply supplanting it (as some energeticists were suggesting).

That then accepted scientific theory could be altered by contemporary researches he taught directly in his textbook, for example, in stipulating the *apparent* independence of electric charge from inertia, and in hedging that, in the present limits of precision, electric masses (charges) were not determined by the quantity of matter that they carry; that without gaining or losing matter, '*au moins de façon appréciable*', a glass rod might have very different electric masses (charges).[101] In intimating or espousing such new ideas, Perrin did not regard himself as a radical in writing *Les Principes*, for 'one can sometimes safeguard the general order of an old edifice, on the condition of accepting numerous and rather profound modifications'.[102]

If certain criticisms were only implied in the section on mechanics, they were much more explicit in Perrin's treatment of the first and second laws of thermodynamics, an exposition ('Le principe d'équivalence et la notion d'énergie') which was also published in the *Revue de Métaphysique et de Morale* (1903). Especially under fire in his discussion of the traditional principle of the conservation of energy were the abuses of the energeticists in the use of the word 'energy' itself. Therefore Perrin's principle is that of *equivalence*, not of the conservation of *energy*, and his interpretation is in the connotation of changes (*changements*), defined by recognition of initial and final states.

In dealing with a change of state from the thermodynamic viewpoint, Perrin grapples first with the problem of causality, as had the critics of mechanics, mechanical models and atomism (for example, Mach).

> From the philosophical point of view, an objection can be made—and it has been made—which hardly appears serious to an experimentalist. I say that the effect of change is recognised in that which would not have taken place in the absence of this change . . . it is not necessary to apply

> unwarily the postulate of causality; and it is not necessary to deduce as certain, from the fact that two systems are practically identical, that their subsequent evolution will be practically the same. . . . But if divergencies are produced, they will obey the principle of equivalence, such as it is soon going to be enunciated: *This is why we will be able to consider this principle as furnishing a form under which causality can be perceived experimentally.*[103]

'Briefly,' he continues, 'in the general case, a change is accompanied by exterior changes to which it is linked by a rigorous causality.'[104] Isolable changes are those realised at the price of an infinitely small exterior change, for example, the play of a snap (*déclic*) or the opening of a stopcock. Tracing historically the kinds of isolable change, Perrin demonstrates that experience with simple machines, calorimetry, and the production of heat by work (thermochemistry) progressed into a belief in the equivalence of different mechanisms for producing changes, such as work and heat. This equivalence permits the association of one measure to all changes and leads Perrin to the conclusion:

> If at the price of a change *C*, one has been able to obtain a change *K*, one never will obtain for the same price, no matter what the mechanism used, first of all the change *K*, and in addition another change, unless the latter should be from that which one knows, by other means, that it costs nothing to produce or destroy.[105]

Perrin follows this statement of the principle of equivalence with definitions and discussions of positive and negative changes, pointing out that just as the appearance of a positive charge in electricity is accompanied by the appearance of a negative charge, so is the presence of a positive change attended by that of a negative change.[106] He then studies in detail three types of systems: a system of weights, a system of thermostats, and a system of dynamos activated by weights plunged in arbitrary thermostats. Using these, he analyses the production of work and the transformation of heat into work through reversible processes.[107] Rather than 'forms' of energy, he considers 'certificates of origin' for the work accomplished, consistently demonstrating that the word 'energy' is unnecessary, that it is not an entity in itself.[108]

As if this were insufficient opposition to the energeticists, Perrin extends his arguments to a redefinition of the second law of thermodynamics, a definition firmly grounded in kinetics and the mechanical models of Maxwell and Boltzmann, rejecting the nebulous and metaphysical phraseology of 'energy' and 'flow' of heat. After reflecting with Paul Langevin on the problem, Perrin promulgates as a first statement of the second law: 'When a change is isolable, the inverse change is not,' or 'An isolated system never passes twice by the same state'.[109]

> Thus, while the first principle is not concerned with knowing, in regard to two phenomena caused by each other, which is *anterior* to the other—which is the *cause* and which is the *effect*—and in some manner ignores all distinction between the future and the past, the second principle affirms a necessary order in the succession of these two phenomena, without a possible return to the states already traversed. This is why I have thought to name this principle a *principle of evolution*.[110]

This idea rests faithful to the meaning of Clausius, according to Perrin, because the Greek word from which Clausius derived the word 'Entropie', signifies 'evolution'.

Subsequently discussing the behaviour of a gas which is first enclosed in one section of a double container and then allowed to diffuse through a stopcock into the second section as well, Perrin points out that it is extremely improbable, and yet not logically impossible, that in their mutual collisions, the molecules assume at a given moment a distribution of velocities which restores the entire ensemble to the first section alone. It will not happen once in one hundred years, and an observer would never see it happen a second time, but it is possible. Hence the second law becomes:

> It is highly improbable that an isolated system passes twice by the same state, it is moreover more improbable as the complexity of the system is greater, and practically speaking, it would be senseless to find a place in this hypothesis for a return to the initial state.[111]

This statement completely coincides with the kinetic theory and probabilistic interpretation of Boltzmann, and it disputes the absolutist interpretation of the second law used by Ostwald, Helm, Mach, Duhem, and (at this time) Planck.

After restating the second law in terms of heat, in order to deduce Clausius' statement and Carnot's principle in the traditional manner, Perrin mathematically derives the entropy quantity. The law can best be expressed, he wrote, as 'The entropy of a system which passes, by an adiabatic transformation, from a first state of equilibrium to a second state of equilibrium is greater in the second state.'[112] The application of this concept to living beings or to systems in explosion is hazardous, he reiterates, while its relevancy for the entire universe involves further problems: it can never really be discussed in a precise sense for other than an isolated system.

Rather, in judging the evolution of the universe, the best assertion that can be made is in the context of the second principle, that '*l'univers ne revet jamais deux fois le même aspect*'.[113] In the current excitement over the death of the universe, it is assumed that the universe is a finite system and that at a certain point in time all isolated systems within it will attain a state of equilibrium, that is, die. Again in agreement with Boltzmann, Perrin cites the kinetic theory and asserts that even if the universe is finite, nothing can prevent a perturbation from occurring in some isolated system, spontaneously disrupting the equilibrium of that system. In addition to this foil to the death of the universe, there are new considerations concerning the evolution of matter itself. If atoms and molecules are not 'êtres morts', as formerly assumed, it seems even more unlikely, suggests Perrin, that there should be an end, or even an interruption, in the life of the universe.[114]

As with the application of the entropy concept to the universe, the successful juxtaposition of a precise theoretical concept against experimental practicability must be a constant brake to metaphysical speculation. In speaking of the concept of reversibility, Perrin assessed that 'perfect rigour is often useless in a science of observation, due to the inherent incertitude in all measurement . . . the preceding definition of [reversibility] is in a way too puristic and hinders us from regarding as reversible certain transformations which have practically the same role, the same utility, the same importance as properly-termed reversible transformations.'[115] Similarly, in treating chemical equilibrium, he adds that despite the fact that [Duhem's] distinction between apparent and real equilibrium is quite precise, it is impossible, in practice, to be absolutely sure that a 'real' equilibrium is not merely apparent and the result of an insufficiently long period of observation.[116]

The energeticists' condemnations of the atomic hypothesis on the grounds of its metaphysical speculation or of its incapacity for experimental proof is seen as not only unacceptable, but as completely hypocritical, for their own theorems lead to the very same problems. Generalising from the various analogies that appear in physical changes to some form of action or 'energy' is especially dangerous, Perrin warns, if it leads to the idea of a mysterious entity which can be divined in all changes of matter or ether, possessing an independent reality as if it were an indestructible and eternal fluid. 'Nothing authorises this vague and obscure hypothesis, which is nevertheless certainly apparent in a number of minds.'[117]

If energeticists accuse atomists of an irresponsible use of hypothesis, obviously the charge can also be levelled at the plaintiffs. Further, as Perrin constantly remarks in *Les Principes*, there are a number of experimental results which establish a firmer reality for corpuscles—atoms and molecules—than one can claim for energy. The experiments of J. J. Thomson and H. A. Lorentz have revived the old, hypothetical theory of electric fluids in an entirely new context: 'Instead of considering electric fluids to be *contained* in matter, it is assumed that they are inseparable—*that they are matter itself.* Thus the material universe will be constructed with only two primordial [electrical] substances.'[118] In addition, one finds with the advent of more advanced experimental technique, a heterogeneity—a discontinuity—reaching further and further into the substrata of apparently homogeneous materials, as demonstrated in microscopic observation of blood and milk, or in colloidal solutions, whose heterogeneity has not yet been observed, but has been strongly suggested by their display of the Tyndall effect (*bleu de Tyndall*).[119] Whereas the energeticist critique of atomism is both philosophical and experimental, the energeticist structure itself is deeply in need of methodological resynthesis; and the advance of experimental technique, Perrin concludes, can only aid the atomists.

That *Les Principes* did indeed succeed as one approach to reviewing and restructuring the controversial aspects of mechanics and thermodynamics can be seen from the extremely cordial reception accorded the new textbook. Wrote Abel Rey in a review, 'it is . . . a lesson of scientific integrity, a model of precision, which the "philosophes" unfortunately need.'[120] Rey compares Perrin's method of generalisation—from a simple, observable case into a precise and clear definition or theory—to the methodology of Greek rationalism or of Renaissance science. 'Let us not forget', he writes, 'that many people are today rejecting it, and are returning in more or less open fashion to the scholastic method, posing arbitrary general laws to which new terms are added for each new phenomenal appearance.' Such was also the verdict of the Société de Philosophie, which devoted a meeting in 1906 to the ideas of *Les Principes*, with Perrin delivering a paper to the group.

This paper, in giving a résumé of the book, differed only in its treatment of the probabilistic interpretation of the second law. Here, new experimental information had come to Perrin's attention since 1903, because of Siedentopf's and Zsigmondy's invention of the ultramicroscope and Perrin's own studies of colloids after his completion of *Les Principes.* As will be seen later, he had read Léon Gouy's papers on Brownian movement and was familiar with the phenomenon's qualitative properties, as well as with Gouy's speculations on its thermodynamic implications. Rhetorically asking if energeticists were correct in sacrificing the molecular hypothesis in order to maintain the absolute character of the principle of Carnot, Perrin reveals that experimental arguments for the uncertainty of that principle do exist.

First discussing the qualitative aspects of Brownian movement as they appear in Gouy's papers—the irregularity of movement, the relation of particle size to movement, the elimination of temperature differentials, external shocks, activation by light, evaporation, etc., as causes of the movement—Perrin asserts that Brownian movement can most easily be explained by the molecular hypothesis and by a probabilistic interpretation of Carnot's principle. As a mechanism to show the uncertainty of Carnot's principle he cites Gouy's example: if one imagines strings attached to each of the Brownian particles, the total pulling of the particles on the strings furnishes ordinary work borrowed from one single source of heat, a source of heat that is eternal in its action. But Perrin grants that such a mechanism rests so far outside our usual way of thinking that some unforeseen difficulty may well exist which would reinstate the absolute nature of Carnot's principle.

Two further possibilities may be imagined which are more troublesome to refute. For example, one may conceive a liquid containing in its lower layers Brownian particles which, due to the perfect irregularity of their movement, expand across the liquid rather than simply resting under a certain level. If a piston formed of a membrane permeable to the liquid, but impermeable to the suspended particles, is placed in the liquid above the particles, their spontaneous diffusion will drive the piston back, performing work in analogy to isothermal expansion of a gas.

On the other hand, if the particles were electrical conductors, for example, particles of gold, one could place them in a magnetic field. While the field is switched off, the particles assume all imaginable distributions and, in particular, one where they all are reassembled in the lower part of the liquid. In the subsequent spontaneous expansion of the particles, within a now active magnetic field, they exercise a force on the magnet which can be utilised for raising weights. This process can be repeated *ad infinitum*. 'Thus the perfect irregularity of Brownian movement leads to the same conclusions as the molecular hypotheses, with the advantage that these conclusions can no longer be easily rejected, for this time they are founded on experiment and direct observation.'[121]

With these mechanisms, based upon experimentally-observed Brownian movement, Perrin visualises no alternative to a statistical interpretation of the second law, an interpretation in keeping with his statement of the law: that an isolated system never passes again by an anterior state. In assessing the relationship of the first and second laws of thermodynamics, Perrin concludes that, '. . . if the principle of equivalence gives an experimental aspect to the principle of causality, the second principle expresses the absolute chance with which molecules and radiations are disseminated in the universe.'[122]

In the discussion at the Société de Philosophie which followed Perrin's paper, several major points were brought up. Paul Painlevé questioned, among other things, whether Perrin's statements of the first and second laws of thermodynamics really involved or necessitated the Carnot–Clausius principle and the idea of evolution in a privileged sense.

> In a general manner, I am not at all convinced by the very profound arguments which tend to make of the Carnot–Clausius principle a *statistical* truth; or at least I believe that one can apply analogous arguments to all other principles (principles of equivalence, of action and reaction, etc.), where it is attempted to portray the elementary mechanism.

To this, Perrin replied that his goal was indeed to instigate those changes which might soon appear necessary in other commonly accepted laws, in consequence of the new molecular interpretations of physical phenomena.[123]

Georges Sorel, also present at the meeting, inquired if Perrin's explication of Brownian movement was in agreement with the new book of Émile Picard, *La Science moderne et son état actuel* (1905), which presented Gouy's analysis of Brownian movement and pointed out that Gouy's experiments were not contrary to Carnot's principle if one applied the principle only to means and to ensembles. Perrin replied that the *visible* movements of Brownian particles were themselves only a mean, and that the interpretations of Picard and himself were identical, requiring that Carnot's principle is less applicable as one deals with more complex systems and that it is never absolute. Similarly, in reply to Painlevé, Perrin insisted that it is artificial to compare a phenomenon as macroscopic as the configurations or evolutions of stars with those phenomena resulting directly from the irregularity of molecular movements.

Of special interest at the meeting was the whole question of the direction of evolution and the inevitable question of the death of the universe. As in *Les Principes*, Perrin replied in complete agreement with the expositions of Boltzmann, that the direction of evolution is towards greater and greater de-coordination of molecular movements. To André Lalande's question whether modern physics justifies the hypothesis of a purely kinetic universe ('conçu à la manière cartesienne'), in which all forms of sensible energy are only corpuscular movements, Perrin answered affirmatively. 'If the kinetic theory of gases has undergone some difficulties, it has on the other hand rendered too many great services to be abandoned.'[124] His opinion regarding the probability of a death of the universe was also in keeping with Boltzmann's, although both Sorel and Lalande were dubious about that interpretation.

The verdict of the group upon Perrin's approach and exposition was expressed by Painlevé. Certainly they did not agree with Perrin upon all points, but the modifications suggested were not at variance with the essential contents of *Les Principes*, and Perrin's answers to their few criticisms were persuasive. Painlevé's opinion was that

> The exposé of Monsieur Perrin appears to me an extremely interesting attempt to make the two fundamental principles of energetics intuitive and to make us grasp their spirit. The objection which can be made to all attempts of this kind is that simplicity can be obtained only at the cost of precision. But I do not believe that one can be more precise than Monsieur Perrin, if one also wants to remain simple.[125]

Painlevé concluded that, regardless of certain minor modifications which he could wish incorporated into the exposition, and which would in no way change its *esprit*, 'the work of Monsieur Perrin appears to me to constitute currently one of the best, if not the best, mode of exposition of the principles of Energetics'.[126]

In Defence of Discontinuity

In addition to his clarification of thermodynamics in *Les Principes*, Perrin most assuredly used the text as a podium for the atomic-molecular hypothesis. He also pursued this goal in the publication of two articles in popular journals in 1901 and 1906. The first of these, 'Les hypothèses moléculaires,' arose from a lecture delivered to students and to the Amis de l'Université de Paris.[127] There he launched a recital of the experimental proofs for the molecular hypothesis, after reviewing the traditionally philosophical character of the postulate, under which partisans of the void debated those of the plenum, with neither daring to dream of experimental verification.

In contemporary endeavours, according to Perrin, present techniques of building microscopes would limit perception of particles to 0·25 micron (1 micron = 0·001 mm). Yet, heterogeneity is verifiable in apparently homogeneous mediums in many other ways. For instance, gold foils of less than 0·25 micron diameter have been made, and only a lack of agility prevents the production of even more fragile ones. From the colour of its reflected light, the black spots (*taches noires*) in soap bubbles have been calculated at 0·01 micron by Reinold and Rücker. In studying oil films on water, Lord Rayleigh has concluded that such films are no more than 0·002 micron thick.

Yet, although one may view an ever finer discontinuity in matter and although one may infer the presence of molecules, the visibility of molecules themselves, continues Perrin, '. . . is not possible at this time'.[128] But *indirect* verifications do exist in the kinetic theory of gases as worked out by Clausius and Maxwell; this places the diameter of molecules between 10^{-7} and 10^{-8} centimetres (suggesting that Rayleigh's oil films contain between two and twenty molecules). It also provides estimates of Avogadro's number, approximating it at $5{\cdot}5 \times 10^{22}$. The law of corresponding states, enunciated by Van der Waals, shows that these same gas laws may be applied to liquids, and the work of Raoult and Arrhenius supplies a new argument for the molecular hypothesis in the theory of dissolved ions.

In cataloguing the characteristics of cathode rays and X-rays, Perrin discerns evidence for an atomic substructure, since cathode particles have a mass 10^{-3} that of a hydrogen atom. 'It is remarkable that the existence of these corpuscules, thanks to the strong electric charges which they carry, is demonstrated in a more direct manner than that of atoms or molecules, which are much larger.'[129] With remarkable imagination, Perrin postulates a structure for the atom (before either Nagaoka's or Rutherford's first hypotheses), so that Louis de Broglie later observed that the famous Rutherford–Bohr atomic model '. . . could very justly be called the model of Jean Perrin'.[130] Perrin suggested,

> Thus we know how to divide an atom into two parts; but the two pieces are not of comparable size, and one of them, the corpuscule, is very small by comparison to the atom. . . . The small piece, the corpuscule, is negatively charged . . . we are forced to say that the larger of the two pieces which proceeds from the rupture of the atom is charged positively . . . the negative corpuscules always appear identical to each other, no matter what the chemical nature of the atom . . . . [Hence] . . . each atom will be constituted, on the one hand, by one or several masses very strongly charged with positive electricity, in the manner of positive suns whose charge will be very superior to that of a corpuscule, and, on the other hand, by a multitude of corpuscules, in the manner of small negative planets, the ensemble of their masses gravitating under the action of electrical forces, and the total negative charge exactly equivalent to the total positive charge, in such a way that the atom is electrically neutral.[131]

Further, Perrin postulated, a cathode ray could be produced when an electric force detaches a negative corpuscle from the atom; subsequently, it would be twice as difficult to detach a second one because of the excess of total positive charge. To explain spontaneous radioactivity, he suggested that in heavier atoms, some of the corpuscules are quite remote and hence not held very tightly: these may easily be severed from the atom.[132]

The fecundity of Perrin's mind is quite apparent, as is its preoccupation with the molecular hypothesis. At the time of publication of this article, Perrin was performing very little experimental work because of the organisation of his course and the writing of *Les Principes*. By the time of his next popular article on the subject, he had aggrandised a prolific background of research and of reading on colloids, and to the proofs for molecular reality assembled in his first article, he added a new world viewed through the ultramicroscope. As he had warned in the 1901 article, further magnification of objects was impossible with the older techniques of microscopy, but the 1903 optical arrangements of Siedentopf and Zsigmondy, using lateral illumination, opened up entirely new horizons.

After reviewing the observations of Brown and Gouy in his 1906 article, Perrin writes that observations of Brownian movement reveal visually the qualitative properties of a fluid in equilibrium:

> This equilibrium exists only in mean fashion and only for large masses: it is a statistical equilibrium. But, in reality, the entire liquid is agitated indefinitely and *spontaneously* in movements all the more violent and rapid as they involve smaller parts and a higher temperature. The observed homogeneity was only apparent: *there is no really homogeneous liquid.*[133]

Rather, all homogeneous bodies are composed of perfectly elastic molecules, animated in rapid and eternal movements, where mean velocity is fixed for a given temperature and increases as the body is warmed up. It is reasonable to allow that the same molecules subsist in solids, as in gases and liquids. The three states of matter differ, then, in the mobility of their molecules and in the frequency with which their molecules change neighbourhoods.

Since apparently continuous matter is formed of molecules, one faces the question of whether there is continuity in the interior of the molecules themselves. The affirmative response to this query, explains Perrin, comes from the chemists who have found it necessary to utilise the notion of the atom. In turn, the facts of radioactivity indicate that the atom itself is a complex structure. Likewise, cathode-ray observations suggest that electricity, like matter, is discontinuous, and that its identical corpuscles are indivisible by all known means. But, Perrin points out, it cannot be said that even these corpuscles fix an 'infranchissable' limit to the subdivision of matter. Traditional notions of continuity result from but an arbitrary choice of our attention among the impressions which are most familiar to us; in reality, discontinuity forms the structure of matter.

Colloid Studies: An Old Problem and New Insights

The colloid studies that assured Perrin of the truth of this statement had been begun several years after his thesis work and were first published in 1903. So he was immersed in colloid literature at the time of one of the greatest breakthroughs in colloid chemistry—the invention of the ultramicroscope. Through these studies, he first became thoroughly acquainted with the phenomenon of Brownian movement and with the new observations of it made with the ultramicroscope.

Perrin's initial interest in colloids arose while he was preparing his lectures for the Sorbonne, when he became confused by Hittorf's explanations of ion velocities in electrolytic solutions. The reference point for these velocities seemed most obscure. For example, to determine the speed of a swimmer in a river, one must assess his velocity with respect to the bank, to a given boat, or to the water of the river. For ions, reasoned Perrin, the velocity must be in relation to 'l'eau du fleuve', that is, with reference to the solvent, and this did not appear to coincide with Hittorf's velocities. Hence, Perrin endeavoured to immobilise the solvent in order to avoid any reflux and change in level produced by the change of concentration at the electrodes. Intercalating a porous diaphragm at one electrode, in hopes of hindering displacement, he suddenly rediscovered the phenomenon of electric osmosis.[134]

What he was really curious about was ion transport and the problem of 'électrisation par contact'—the electric charge on walls washed by an electrolyte—but he perceived in electric osmosis a simple means of studying his principal topic. Accordingly, his first experiments with electric osmosis, using porous membranes of graphite, sulphur, or napthalene in aqueous solutions, led him to the definition of several rules: (1) that electric osmosis gives an easy means of studying the charge of contact between an ordinary solid and a liquid; (2) that this charge is, on the average, much larger when the liquid is a good ionising agent, such as water, and is due to ions present in the liquid; (3) that in water, H^+ and OH^- are the only ions extremely active; each charges the wall with its sign; and (4) that all positive polyvalent ions diminish the action of OH^- ions, while negative polyvalent ions diminish the action of H^+ ions. This paralysing action increases with concentration and valence.[135]

Perrin believed the charge on the wall to be compensated by the charge of a thin layer of liquid near the wall. The nature of the wall seemed to play only a secondary role, and this intimated that there should be a double electric layer at the surface of water in contact with air or a gas; the direction of this double layer, and its importance, should depend on the presence of acids, alkalis or polyvalent radicals. Such electrisation should also lower the surface tension. Indeed, measurements proved this to be true.[136]

The idea that colloid particles were charged electrically had pervaded the colloid literature for some time and the experiments of Cotton and Mouton, detecting the movement of colloid granules in an electric field, had established this point. In his papers, Perrin cited the opinion of Hardy (of the Cavendish laboratory) that the electrisation of the solid colloid granules by contact with water was a necessary condition of stability of a hydrosol. Surveying the colloid data of the last few years, Perrin also knew that the sign of this charge is often extremely sensitive to a slight excess of acid or base and that colloids are coagulated by the addition of electrolytes. This coagulation is of greatest magnitude when the added electrolyte contains polyvalent ions of sign opposite to the colloid and when the valence of these polyvalent ions is highest. Further, the granules have no charge at the moment of coagulation.[137]

Perrin then drew a parallel between the rules of electrisation by contact and those of colloid behaviour: 'La symétrie de ces énoncés est remarquable.'[138] His rules could explain in a more encompassing theory how an equilibrium of colloid granules of a given size is reached, why the granules reversibly enlarge or diminish when there is a change of concentration in the liquid, and why an irreversible coagulation is produced when the addition of electrolytes, especially of polyvalent ions, exceeds certain limits.

> Briefly, it is necessary to indicate the causes which assure a *stable* equilibrium for a certain diameter of the granule. Towards this goal, I will propose a theory, which is summed up in the following sentence: *Surface tension and cohesion favour the enlargement of a granulèe, but the electrisation of the granule is an internal cause of dislocation and one conceives that there exists a diameter for which these two opposite influences are in equilibrium.*[139]

That is, when a germ of a granule reaches a certain size, it may carry one electron; when it becomes larger, it may carry two, but repulsion of the electrons may then cause a fragmentation of the larger particle. It follows that if one increases the electrisation (by addition of the proper acid or base), one diminishes the mean size of the granule corresponding to equilibrium. If one decreases the electrisation, one increases this size. If one decreases the charge beyond a certain critical value, segmentation becomes impossible, and the granules are aggregated by their surface tension: this is coagulation.[140]

Perrin's results were a major contribution to the field of colloid research, although when he first began he had no idea of the incredibly large amount of literature already existing on the subject. His approach to the problem was typical, however, of his general procedure in experimental work. By and large, Perrin undertook research without studying the bibliography and detailed literature on a subject. He felt that by looking too completely at other papers, one might become bogged down in others' dead ends or perhaps become convinced that there was nothing further to do. For this reason he looked at extended bibliography only after his initial experiments. With colloids, for example, he interpolated a barrier in an ionic solution, observed the results, and subsequently discovered that this had been done fifty years earlier.[141] His attack on the cathode-ray problem was somewhat similar, if one remembers that he was unaware of the experiments of Hertz, and, as will soon be discussed, the same incomplete exploration of the literature occurred with his initial experimental work in Brownian movement.

At least two other general aspects of Perrin's approach in solving a problem can be seen in his studies of electrisation by contact—first, his documentation of the history of a topic and, secondly, his application of the total results in as universal a fashion as possible. In the electrisation studies, for example, he first published several short papers in the *Comptes rendus* and then recapped them in a lengthy paper of 1904–1905, prefaced by a short history of investigations in electric osmosis and electrisation. This procedure can also be traced in his work on cathode rays and X-rays, where his final doctoral thesis was prefaced by an historical discussion. This will be seen, too, in his studies of Brownian movement. His emphasis on the history of a problem is in part explained in a critical book review which he wrote in 1907, where he noted that in the profound transformation which physics was undergoing, those preoccupied with the revolution rarely wrote its history, leaving those who undertook that task quite free to paint too nice a picture of the birthpangs of a theory.[142] Perrin hoped to avert this whitewashing.

To see his extension of specific experimental results to a far broader problem, one has only to turn to his own statement regarding his experiments on cathode rays, X-rays, ion transport, colloids and Brownian movement: 'My works have been especially of the experimental order, and attached to Atomistics.' He was ever conscious of marshalling his results for the proof of molecular reality. In addition, he was particularly mindful of the biological consequences of physico-chemical laws. Electrisation by contact attracted his interest because it played 'a capital role in the properties of colloids and consequently in the properties of the granules or *micelles* which appear to form all living matter'.[143] Indeed, in terminating his papers on the subject in 1905, he postulated that the formation of chromosomes might be due to an increase of electric charge, according to rules which he had defined.[144]

In the final paper on electrisation by contact, one sees Perrin's intrigue with the phenomenon of Brownian movement. As mentioned earlier, the only major French physicist who had dealt with this problem was Léon Gouy, and the phenomenon excited Perrin's attention through his own personal observation and through the colloid literature. As he was later to reiterate in his popular 1906 article in *Revue du Mois*, 'the discontinuous structure of colloid solutions should now be considered as a reality furnished directly by experience and no longer as a hypothesis'.[145] He himself originally observed the Brownian movement with a regular Abbe microscope; with the ultramicroscope, he gazed at it in even more minute dimensions. From his first observations of the general properties of the granules, he immediately culled imaginative speculations concerning their place in the general kinetic theory, which had henceforth only been applied to gases and dilute solutions.

With Paul Langevin to advise him, Perrin quickly attempted kinetic energy calculations for the Brownian granules. Langevin reasoned that the mean kinetic energy of a colloidal particle should be less than that of an isolated molecule. Using gross calculations, Perrin found a mean energy 100 000 times smaller than that assigned by the kinetic theory to a gas or liquid molecule at the same temperature.[146] He further carried out freezing point and vapour pressure measurements, hoping to find a correlation with Raoult's laws, but again obtained unsatisfactory results, which merely indicated impurities in solution.

> It can be objected that, due to the fact of their Brownian movement, the granules which form the colloid exercise a certain osmotic pressure on each wall which stops them. Then, if the reasoning of van't Hoff remains applicable, there will be a lowering of the freezing point, due to the granules themselves and not to impurities.

Yet, with his figures for the mean kinetic energy of the granules and for their number, the values for osmotic pressure and lowering of the freezing point fell well below those which appeared feasible.[147]

A third manner in which the Brownian granules might be directly linked with general kinetic theory was in their diffusion, which should be a consequence of their irregular displacements. Perrin read a discussion of this thesis in the 1904 doctoral dissertation of Jacques Duclaux (who subsequently devoted most of his research to colloids), and after considering the implications of such a diffusion, Perrin postulated,

> In reflecting on this diffusion, one sees above all that it cannot have the effect of rigorously equalising the distribution of the granules. It is only in a horizontal plane that this distribution becomes and remains uniform. But the concentration of the granules should increase as this plane is taken at a lower level (if the granules are at least more dense than the liquid in which they are immersed).
>
> *Thus, the diffusion has the effect of establishing, then of maintaining invariable, a certain mean distribution of granules, such that a certain element of volume contains all the more granules as it is at a greater depth in the liquid.*
>
> It is thereby explained, for example, that a colloidal solution of copper ferrocyanide, which is left alone for a long time, should always be darker in the lower part, *and this is by no means the index of a precipitation.* If the intensity of weight was, for example, a thousand times larger, the inequality of equilibrium distribution would be even more marked. That explains this observation, given without interpretation by Jacques Duclaux, according to which, in centrifuging a solution of copper ferrocyanide several hours, the upper layers can be almost completely discolored.

In addition, remarked Perrin, this continuous increase of concentration with depth is not a property separating colloidal solutions from true solutions. On the contrary, Gouy had shown by thermodynamical reasoning that a solution of copper sulphate should be more concentrated at lower levels than at higher ones. Further, in the Pays-Bas, Lobry de Bruyn produced and measured similar differences of concentration between the upper and lower layers of a centrifuged homogeneous solution.[148]

Though there is no mention here of the older Laplace function, which gives an exponential distribution for the decrease of gas (atmospheric) concentration with height, Perrin's mind is clearly proceeding towards a profitable experimental study of Brownian movement, which could result in a more exact calculation of the fundamental constants of the molecular-kinetic theory of matter—such as molecular diameter, Avogadro's number, and the fundamental electric charge. Thus far, only indirect methods of determining these constants had been implemented, such as calculations based on gas viscosities. Relative values had also been found, as in Thomson's e/m determinations. The phenomenon of Brownian movement presented, for the first time, the possibility of direct and absolute measurements of visible particles appearing to conform beautifully to the tenets of the kinetic theory. If values derived from these granules coincided with figures calculated from indirect and relative methods, the foundations of the kinetic theory would rest assured, as would the absolute truth of molecular reality. The taskmaster for Perrin's further research was clearly beckoning him.

NOTES

1. Max Planck, 'Scientific Ideas, Their Origins and Effects', in *The New Science*, trans. James Murphy (New York: Meridian, 1959), pp. 298–299.
2. Carl L. Becker, *The Heavenly City of the Eighteenth-Century Philosophers* (New Haven: Yale University Press, 1932).
3. Louis de Broglie, 'Contemporary Science and Traditional Values', pp. 214–217 in *New Perspectives in Physics*, trans. A. J. Pomerans (Edinburgh, 1962), p. 216.
4. Francis Perrin, 'Jean Perrin, prix Nobel de Physique—1926', unpublished. A copy of this manuscript was kindly given to me by M. Perrin in Paris, autumn 1968.
5. Albert Ranc, *Jean Perrin, un savant au service du socialisme* (Paris: Editions de la Liberté, 1945), pp. 20–21.
6. See *Oeuvres Scientifiques de Jean Perrin* (Paris: Centre Nationale de la Recherche Scientifique, 1950), p. xii, for a list of the professional memberships and honours which Perrin held. This volume (hereinafter referred to as *Oeuvres Scientifiques*) contains thirty-four of Perrin's published papers as well as an almost complete bibliography of his papers (113), books (5) and brochures (15).
7. Elsa Thulin, *Souvenirs sur Jean Perrin*, (Stockholm: Institut Français de Stockholm, 1949), p. 4.
8. Interestingly enough, Francis Perrin also studied under Marcel Brillouin, though at the Collège de France rather than at the École normale. His earlier education was much different from his father's, however, as he attended no schools until the family was in New York, where he was sent to the Horace Mann School for three months. Like his father, he was interested in science and mathematics from childhood, and at age thirteen, Francis was performing mathematical integrations under the at-home tutorship of family friend Émile Borel. In a conversation in Paris, 29th October, 1968, M. Perrin said that his mother taught him to read and write, while his father and father's friends directed the rest of his education.
9. In the chapter on acoustics in Jean Perrin's personal copy of his *Notice sur les Travaux Scientifiques* (Toulouse: Edouard Privot, 1923), there are numerous additions in pencil, notations of figures, and deletions of the printed text, indicating that Perrin rewrote this chapter into the 1932 article.
10. Jean Perrin, 'La fluorescence', *Annales de physique*, **10** (1918), p. 133.
11. Francis Perrin, 'L'oeuvre scientifique de Jean Perrin', *Institut de France, Académie des Sciences* (*Publications Diverses*), #14 (1962), p. 18.

12. Jean Perrin, 'L'origine de la chaleur solaire', *Scientia*, 1921, p. 345; in *Oeuvres Scientifiques de Jean Perrin*, p. 385.
13. Ernest Rutherford, 'La structure de l'atome', pp. 36–64 in *Atomes et Électrons* (Paris: Gauthier-Villars, 1923). In this paper, delivered to the Council of Physics, Rutherford discussed the experiments of Moseley and of Geiger and Marsden, Darwin's calculations, Bohr's spectral studies, dimensions of the atomic nucleus, artificial disintegration, and isotopes.
14. *Atomes et Électrons* (1923), p. 68. This remark was made in the discussion following Rutherford's paper.
15. Ibid., p. 75.
16. Reino Virtanen, 'Marcelin Berthelot. A Study of a Scientist's Public Role', *University of Nebraska Studies*, new series #31 (April, 1965), p. 22.
17. Ibid., p. 24.
18. Ibid., pp. 37–48.
19. *Livre d'hommage des lettres françaises à Émile Zola* (Paris: Société libre d'Édition des Gens de Lettres, 1898), p. 33.
20. Léon Blum, preface to Ranc's *Jean Perrin, un grand savant au service du socialisme*, p. 16.
21. Léon Blum, 'Jean Perrin', preface to the collection of excerpts from Jean Perrin's publications, *La Science et L'Espérance*(Paris: Presses Universitaires de France, 1948), p. xxxi.
22. Ibid., p. xxiv.
23. Francis Perrin, 'L'oeuvre scientifique de Jean Perrin', p. 6.
24. Jean Perrin, 'H. G. Wells', in *La Science et L'Espérance*, pp. 115–116.
25. Jean Perrin, 'La science liberara l'homme', in *La Science et L'Espérance*, p. 177.
26. Ranc, p. 39.
27. Henri Laugier, in a contribution to the *Hommage National à Paul Langevin et Jean Perrin* (Paris: Orléans, 1948), pp. 33–34.
28. Léon Blum, 'Jean Perrin', p. xxxi.
29. As the C.N.R.S. began its work, its registry numbered about 40 researchers, 3 million francs gold, the Office of Inventions at Bellevue and two or three laboratories. In 1962, its staff included almost 4000 researchers and 4500 technicians, 150 million francs gold, and 100 laboratories or institutions. A first fruit of its development was Frédéric Joliot, who progressed through the 1930's as research scholar, *chargé de recherches*, and master of research, and who made the discovery of artificial radioactivity at the old Laboratoire de synthèse at Ivry with Irène Joliot-Curie. For a fairly complete account of the development of the C.N.R.S., see Ranc; and also, Jean Coulomb, 'Jean Perrin fondateur de Centre Nationale de la Recherche Scientifique', *Institut de France, Académie des Sciences (Publications Diverses)*, #14 (1962).
30. Pierre Girard, in Girard and Paul Lebeau, *Hommage à Jean Perrin* (Paris: Hermann, 1950), p. 23.
31. Jean Perrin, 'Le Palais de la Découverte', in *La Science et l'Espérance*, pp. 151–152.
32. Opened the 24th May, 1937, the Palais hosted two million visitors by the following November. In 1960, 2000 groups of school children watched its demonstra-

tions and displays, according to the figures of Ferdinand Lot, *Jean Perrin et les atomes* (Paris: Seghers, 1963), pp. 59, 76.

33. Ranc, p. 53.
34. Girard, p. 24.
35. Émile Borel, 'Transfert des cendres de Jean Perrin', *Institut de France, Académie des Sciences*, 11 (1948), #p. 2.
36. Louis de Broglie, 'The Varied Aims of Scientific Research', pp. 198–202 in *New Perspectives in Physics*, p. 201.
37. Louis de Broglie, 'Jean Perrin', p. ix. Perrin later married Duportal's sister.
38. Ranc, p. 29.
39. Ibid.
40. Ibid., pp. 20–30.
41. Borel, p. 3.
42. All of Perrin's friends, save one, supported his ideas. The one exception was Georges Urbain who never accepted the general ideas of Perrin. According to M. Francis Perrin, Urbain said that he would accept the concept of isotopes only when separate samples of Cl^{35} and Cl^{37} were placed in his hands. He died in 1938.
43. Paul Lebeau, in Lebeau and Pierre Girard, *Hommage à Jean Perrin*, p. 5.
44. Harry W. Paul, 'The Debate over the Bankruptcy of Science in 1895', *French Historical Studies*, 5, #3 (1968), p. 314.
45. Virtanen, p. 33.
46. Jean Perrin, 'L'astronomie, libératrice de la raison humain', preface to Paul Couderc's *L'Architecture de l'Univers* (Paris: Gauthier-Villars, 1930); in *La Science et L'Espérance*, p. 3.
47. Francis Perrin, 'L'oeuvre scientifique de Jean Perrin', p. 7.
48. Jean Perrin, 'La recherche scientifique', *Actualités Scientifiques et Industrielles*, #58 (Paris: Hermann, 1933), pp. 16–17.
49. Virtanen, p. 46.
50. Jean Perrin, 'La chimie physique', in *L'Orientation Actuelle des Sciences* (Paris: Félix Alcan, 1930), p. 18.
51. Claude Bernard, *An Introduction to the Study of Experimental Medicine*, trans. Henry C. Green (New York: Dover, 1957), pp. 32, 35.
52. *Jubilé de M. Marcel Brillouin. Mémoires originaux offerts à M. Marcel Brillouin à l'occasion de son 18ᵉ anniversaire* (Paris: Gauthier-Villars, 1935), introduction.
53. *Recherches récentes sur diverses questions d'hydrodynamique* (Paris: Gauthier-Villars, 1891), p. 44.
54. Marcel Brillouin, introduction to Ludwig Boltzmann's *Leçons sur la Théorie des Gaz*, trans. A. Gallotti (2 vols.; Paris: Gauthier-Villars, 1902–1903), Vol. I, pp. xiv–xv.
55. Marcel Brillouin, preface to Boltzmann's *Leçons sur la Théorie des Gaz*, p. xviii.
56. Ibid., p. xvii.
57. Jean Perrin, *Notice sur les Titres et Travaux Scientifiques* (Paris: Gauthier-Villars, 1918), p. 8.
58. Girard, p. 15.
59. Jean Perrin, 'H. G. Wells', p. 114.

60. Jean Perrin, 'Les progrès de la Science', preface to Perrin's *Les Éléments de la Physique* (Paris: Albin-Michel, 1929), in *La Science et L'Espérance*, p. 65.
61. Marcel Courtines, 'La Lumière, principe du monde à propos de Jean Perrin', *Cahiers de la Quinzaine*, 18ᵉ series, **4** (Paris: L'Artisan du Livre, 1927), p. 10.
62. For a résumé of the studies of cathode rays, see chapters I. and II. in G. K. T. Conn and H. D. Turner, *The Evolution of the Nuclear Atom* (New York: American Elsevier, 1965).
63. As shown in 1897 by J. J. Thomson, Hertz' conclusions were misleading due to the inefficiency of evacuating techniques at the time. Gas remaining in Hertz' tube was ionised by the cathode rays, neutralising any effects that external electrified plates would have had on the cathode rays.
64. Wrote Perrin in his *Notice sur les Travaux Scientifiques* (1923), p. 14, 'Heureusement, j'ignorais à ce moment que Hertz avait cru établir ce point.'
65. Jean Perrin, 'Nouvelles propriétés des rayons cathodiques', *Comptes rendus*, **121** (1895), p. 1130. Also, in *Oeuvres Scientifiques*, p. 4.
66. Perrin, *Oeuvres Scientifiques*, p. 4.
67. Ibid.
68. Jean Perrin, 'Rayons cathodiques et rayons de Röntgen', *Annales de Chimie et Physique*, **11** (1897). *Thèse de doctorate soutenue 14 Juin 1897*. In the *Oeuvres Scientifiques de Jean Perrin*, p. 14, footnote 2.
69. Ibid., p. 17.
70. *Notice sur les Travaux Scientifiques* (1923), p. 15.
71. Jean Perrin, *thèse* in the *Oeuvres Scientifiques*, p. 33.
72. Ibid., p. 39.
73. Louis de Broglie writes in his preface to *La Science et L'Espérance*, p. xi., that Perrin '. . . was one of the first to realise that, if X-rays make gases conductors or discharge charged bodies, it is due to the creation of ions or to the provocation of the emission of electrons'.
74. Girard, pp. 11–12.
75. Ibid., p. 17.
76. Borel, p. 4.
77. Lot, pp. 96–97.
78. Francis Perrin, 'L'oeuvre scientifique de Jean Perrin', p. 10.
79. Edmond Bauer, in a contribution to the *Hommage National à Paul Langevin et Jean Perrin*, p. 121.
80. Pierre Girard writes, p. 10, that '. . . it is much less for the *candidats à la Licence* and *à l'agrégation* that these lessons are arranged than for the researchers who have already vaulted the successive hurdles of examinations and competitions.'
81. Bauer, p. 21.
82. Ibid.
83. Girard, pp. 10–11.
84. George Sarton, 'Paul, Jules and Marie Tannery', *Isis*, **38** (1947), p. 43.
85. Jean Perrin, 'Les progrès de la science', p. 60.
86. Jean Perrin, preface to *Les Principes. Traité de Chimie Physique* (Paris: Gauthier-Villars, 1903).
87. Ibid., p. xvi.

88. Ibid.
89. Ibid., p. x.
90. Ibid., p. xi.
91. Ibid., pp. xxi–xxii.
92. Jean Perrin, introduction to *Les Principes*, pp. xvii–xxii.
93. Ibid., p. xxv.
94. *Les Principes*, p. 198.
95. Ludwig Boltzmann, 'Über die Grundprinzipien und Grundgleichungen der Mechanik' (Clark University, 1899), pp. 253–307 in *Populäre Schriften*, pp. 257–258; also quoted in Dugas, p. 32.
96. Jean Perrin, *Les Principes*, p. 13.
97. Ibid., p. 30.
98. Ibid., p. 119.
99. Ibid., p. 19.
100. Ibid., p. 6.
101. Ibid., p. 23.
102. Ibid., preface, p. xi.
103. Ibid., p. 76.
104. Ibid., p. 79.
105. Ibid., p. 92.
106. Ibid., pp. 107–108.
107. Ibid., p. 117.
108. Ibid., p. 121.
109. Ibid., p. 142.
110. Ibid., pp. 142–143.
111. Ibid., pp. 145–146.
112. Ibid., p. 178.
113. Ibid., p. 179.
114. Ibid., pp. 179–180.
115. Ibid., p. 36.
116. Ibid., p. 139.
117. Ibid., p. 82.
118. Ibid., p. 23.
119. Ibid., p. 203.
120. Abel Rey, 'Les principes philosophiques de la chimie physique', *Revue philosophique de la France et de l'Étranger*, **57** (1904), p. 394.
121. Jean Perrin, 'Le contenu essentiel des principes de la thermodynamique', *Bulletin de Société de Philosophie*, **6** (1906); in the *Oeuvres Scientifiques de Jean Perrin*, p. 71.
122. Ibid., pp. 71–72.
123. Ibid., p. 75.
124. Ibid., p. 77.
125. Ibid., p. 72.
126. Ibid., p. 75.
127. Jean Perrin, 'Les hypothèses moléculaires', *Revue scientifique*, **15** (1901), 449–461.
128. Ibid., p. 451.
129. Ibid., p. 460.

130. Louis de Broglie, preface to *La Science et L'Espérance*, p. xx.
131. Jean Perrin, 'Les hypothèses moléculaires', p. 460.
132. Rutherford later commented that he was unaware of Perrin's hypothesis when he first published his own version, which, unlike Perrin's, was firmly grounded in experimental work.
133. Jean Perrin, 'La discontinuité de la matière', *Revue du Mois*, **1** (janv.—juin 1906), p. 335.
134. *Notice sur les Titres et Travaux Scientifiques* (1918), p. 29.
135. Jean Perrin, 'Conditions qui déterminent le sign et la grandeur de l'électrisation par contact (III)', *Comptes rendus*, **137** (1903), p. 514.
136. *Notice sur les Titres et Travaux Scientifiques* (1918), pp. 37–38.
137. Jean Perrin, 'Conditions qui déterminent le sens et la grandeur de l'électrisation par contact', *Comptes rendus*, **137** (1903), p. 514; and 'Mécanisme de l'électrisation de contact et solutions colloïdales', *Journal de Chimie physique*, **3** (1905), p. 75.
138. Jean Perrin, 'Mécanisme de l'électrisation de contact . . .', (1905), p. 87.
139. Jean Perrin, 'Électrisation par contact (IV) et théories des solutions colloïdales', *Comptes rendus*, **137** (1903), p. 564.
140. Ibid., pp. 565–566.
141. This was corroborated by M. Francis Perrin.
142. Jean Perrin, 'À propos de "L'Évolution des Forces" ', *Revue du Mois*, **4** (juillet-decembre, 1907), p. 606.
143. Jean Perrin, 'Examen des conditions qui détermine le signe et la grandeur de l'osmose électrique et de l'électrisation par contact', *Comptes rendus*, **136** (1903), 1388–1391.
144. Jean Perrin, 'Mécanisme de l'électrisation de contact . . .', (1905), p. 108.
145. Ibid., p. 51.
146. One can only assume that the velocity figures which he used were those of Exner (1900) or of Siedentopf and Zsigmondy (1903), all of which were too small.
147. Jean Perrin, 'Mécanisme de l'électrisation de contact . . .', (1905), p. 65.
148. Ibid., pp. 60–61.

Chapter Three

The Essential Debate

THE exponential law which Perrin announced in 1908, describing the vertical distribution of a colloid at equilibrium, was the fruit of laborious experiments on Brownian movement after several years of apprenticeship in the study of colloids. Included in his first 1908 paper on Brownian movement was a successful application of the concepts of osmotic pressure and mean kinetic energy to the visible Brownian particles, as well as a convincing calculation of Avogadro's number. These endeavours were but the prelude to a five-year drama devoted to the erection of an unassailable edifice to house the dictum of molecular reality, a structure buttressed at its most vulnerable point of criticism by the observed laws of visible Brownian movement.

Initially Perrin published four papers in the *Comptes rendus* of the Académie des Sciences. All four were delivered in 1908 and were completed with little attention to the theoretical work undertaken a year or two earlier by Albert Einstein and Maryan Smoluchowski. Once Perrin turned to their derivations, however, he realised full well that the strictly experimental work of his papers beautifully confirmed their theoretical predictions, as well as those of his friend Paul Langevin, and he applied anew their equations to other aspects of Brownian movement, utilising the masterful experimental techniques which he had perfected.[1] Although several other chemists, notably The Svedberg and Max Seddig, had attempted to relate the new mathematical concepts to Brownian movement, their results were neither as accurate nor as convincing as those of Perrin, and, as will be seen, they themselves readily conceded the superiority of his work. Certain other efforts, especially those of Victor Henri, originally cast strong aspersions on the theory of Brownian movement, and, as will be discussed later, caused a general doubt about the accuracy of Einstein's assumptions—a doubt which was dispersed only when Perrin triumphantly presented his measurements of the Brownian movement of translation and rotation.

The initial work done by Perrin arose totally from his own experience with colloids, his familiarity with the tenets of kinetic theory, and his determination to discover an irrefutable proof for molecular reality—a purpose thoroughly demonstrated in his earlier researches and writings. A number of colloid chemists had already analysed Brownian movement in colloids, but, as will be seen, Perrin re-oriented such experimental work into another direction which proved quite profitable both to colloid chemistry in general and, more particularly, to the principal question in which he was interested. If his work was beneficial to those scientists interested exclusively in colloids and their behaviour, Perrin was also indebted to them for the questions which they elucidated and for their removal of certain obstacles which might have hindered his own interpretation of the cause of Brownian movement.

Colloid Theories of Brownian Movement

Colloids had originally been recognised, or at least named, by Graham (1861), who separated crystalloids and colloids by their rates of diffusion through a parchment membrane. Arrhenius later made more accurate measurements, and osmotic pressures were determined by Pfeffer in 1877. By the early twentieth century, it was shown that electrolytes which will precipitate a colloid, also diminish the osmotic pressure, and a dependence of osmotic pressure on temperature was also noted. It was accepted, then, by the end of the first decade of the twentieth century that the colloidal particles in some fashion cause the observed pressure.[2]

Besides studies in diffusion and osmotic pressure, several other questions occupied the colloid chemists, especially problems involving the migration of a colloid in an electrical field, the rate of this migration, and the relationship of electrical effects to the phenomenon of coagulation. Under the influence of a potential drop, almost all colloids migrated either to the cathode or anode. The rate of this migration could be measured in a U-shaped tube, where electrodes were immersed in water in the upper part of the two arms, and the boundary between the colloid and the water was followed as it moved either toward the cathode or anode.[3] The existence of electrical effects was explained by a variety of theories. For example, the electric double layer theory supported by Perrin was also favoured by Aimé Cotton and Henri Mouton in France and by E. F. Burton in England. Jacques Duclaux, who became a critic of Perrin in 1908, assumed that charged particles were present, while other viewpoints postulated an absorption of ions or a dissociation of molecules on the surface of the colloid particles. There was also the suggestion that both contact electrisation and an ionisation phenomenon could account for the various observations, the former for coarser suspensions and the latter for better-defined colloid solutions.[4]

There were, of course, a number of inquiries into the size of the colloid particles, with the Tyndall effect implying the presence of a spacial discontinuity even when it could not be observed with an ordinary microscope. The Tyndall effect predicted that the plane polarisation of light by small particles becomes more pronounced with diminishing size of the particles and that it is complete when their diameter is small compared with the wavelength of light. The formulas of Rayleigh related the intensity of light to the particle size, enabling rough estimates to be made of the dimensions of colloid particles.[5] In most suspensions and in turbid solutions, the individual particles could be identified under an ordinary microscope, but in the majority of hydrosols and other colloidal solutions, visibility was impossible. The difficulty lay in the fact that the resolving power of objectives is fixed at about 0·25 micron, and below that limit the light is diffracted by the elements structuring the particle, so that no measurements can be made.[6]

One can imagine the excitement, then, when Siedentopf and Zisgmondy published in the *Annalen der Physik* in 1903 the description of a new apparatus which delved into the structure of matter to the limits of 5×10^{-3} micron. Their principle of illumination was identical to that of the visibility of stars at night. By illuminating the particles with a cone of light at right angles to the light, which is, in turn, diffracted by the particles upwards into a microscope, one might see the polarised diffraction discs as images. Further, in order to avoid the confusing appearance of diffusion discs from particles in the layers above and below the focused layer, one illuminates only those particles which are to be made visible, by focusing the light source through a small spectroscopic slit and an assemblage of telescopic lenses and collimators upon the desired particles in a layer of predetermined thickness.[7] With such a system, the magnification of objects can be pushed to 150 000 diameters (although molecules themselves remain invisible), and at such limits Siedentopf and Zsigmondy viewed a Brownian movement ever more lively the more deeply they peered into the structure of matter.

A modification of this apparatus was immediately developed by Cotton and Mouton (of the École normale and the Institut Pasteur, respectively). In their device, the light source impinges obliquely upon the preparation to be observed, striking the glass at such an angle as to undergo total reflection from the upper surface of the object-cover in contact with air. Thereby the light beam is turned below and does not penetrate directly into the microscope, so that the field remains obscure. The small particles of a colloid or other preparation diffract the light to all sides and particularly into the microscope. Such a system could not be used for immersion studies, but was simpler to assemble than that of Siedentopf and Zsigmondy.[8]

Once the ultramicroscope was developed and a few observations were made, a vigorous interest in investigating the sizes and velocities of the colloid particles arose. Among the most accurate studies completed before the advent of the ultramicroscope had been the researches of Felix Exner, published in 1900. Observing an emulsion of gamboge with an ordinary microscope and an Abbe marking apparatus,* Exner had found velocities of approximately 3–4 microns per second for particles of diameter 0·4–1·3 micron; he had also plotted the temperature-velocity relationship for particles of a diameter 0·7 micron. In attempting to photograph the particles against a dark background, however, the photographic plates proved too insensitive. Exner endeavoured to measure smaller particles, but failed because the ordinary microscope would reveal only what he termed 'bent pictures' (*Beugungsbilder*)—dark circular rings with bright light inside.[9]

Siedentopf and Zsigmondy immediately realised the impact that the ultramicroscope would have upon the classical studies of Brownian movement, such as the relationship between size and velocity of the particles, the effects of temperature, and the questions of a connection with electric charge, capillarity, evaporation or surface tension. After discussing immersion objectives, holders for liquids, and other pro-

* Exner placed the microscope in a double-walled metal box with hot water flowing through lead pipes, insuring an even and constant temperature in the box. To estimate the size of the particles, he marked their path on a smoked-glass plate with a needle, using divisions on the ocular for reference. A picture of the marks was then projected on a screen, the path measured and divided by the magnifications of the microscope and projection lens and by the time of observation.

cedures in their first published description of the new method of observation, they remarked that this method was especially suitable for studies of the 'so-called Brownian movement in liquids' ('. . . dass die Methode zum Studien der sogenannten Brownschen Bewegung im Flüssigkeiten besonders geeignet ist, mag hier nicht unerwähnt bleiben').[10] Zsigmondy carefully studied the motions of minute particles of gold, observing that particles of a diameter 10 millimicrons covered a straight line of 10 microns distance in $\frac{1}{8}$ second, that particles of 75 millimicrons diameter effected movements whose amplitude was approximately 2–3 microns in $\frac{1}{8}$ second.[11]

Exner had concluded in 1900 that the kinetic theory was an insufficient explanation of Brownian movement because the kinetic energies calculated from the observed velocities were neither proportional to the absolute temperature nor identical to those generally assumed for molecules in the liquid state.[12] Further efforts to calculate accurate velocities and kinetic energies with the ultramicroscope eventually led to the recognition that the apparent mean velocities of the particles varied in size and direction, depending upon the length of the time of observation, without tending toward a limit. Although this had never been realised until the employment of the ultramicroscope, its comprehension seemed to raise an insurmountable obstacle to any definitive calculations which might possibly relate the cause of Brownian movement to the kinetic theory. On the other hand, such an observationally illimitable motion convinced others of that very hypothesis, since such a phenomenon only implied particle velocities so great, that an observer —no matter what his apparatus—could never perceive the actual paths traversed, but would instead see a certain mean velocity as an instantaneous velocity. Such a conception was completely harmonious with statistical interpretations of the kinetic-molecular theory.

Cotton and Mouton, for example, were convinced of the applicability of the kinetic theory in explaining Brownian movement. In discussing its appropriateness as at least a partial explanation, they pointed out that kinetic theory indicates a mass of about 18×10^{-24} grammes for a water molecule, while the mass of the gold particles studied by Zsigmondy approximates 2×10^{-14} grammes (for particles of diameter 10 millimicrons). The mass of gamboge particles—commonly used in investigations of Brownian movement—would then be more than a million times larger than water molecules. If one assumes that water molecules are ordinarily 0·5 millimicrons apart, with mean velocities of several hundred metres per second, they would certainly be capable of producing a Brownian movement by multiple collisions.[13]

Still, colloid chemists were not convinced that kinetic theory provided a total explication of the phenomenon. Zsigmondy himself believed that an explanation which would eventually solve the riddle of the motions of colloid particles and Brownian movement must eventually entail some combination of electrical-charge effects and the kinetic theory of liquids, and Cotton and Mouton more or less concurred. Jacques Duclaux,[14] who was to be one of the few critics of Perrin's initial work on Brownian movement, insisted from the very inception of his own experimental researches, that the osmotic pressure of colloidal solutions and the movements of their particles were due to an electrostatic repulsion of charges. He also vehemently maintained that coagulation was due to a decrease of Brownian movement after a large accumulation of electric charges around each particle had enfeebled an electric field originally surrounding the Brownian granules.[15]

Perrin's Experimental Height Distribution

When Perrin seized upon Brownian movement as a phenomenon for detailed study, many mysteries had been illuminated, though irksome questions still remained. There was in the current colloid literature a renewed interest in the Brownian particles as a thing in themselves, with opinion still divided over the cause of their activity. The movement's eternal nature, its existence among all microscopic particles, and the relationship of the granular nature of colloids to an osmotic pressure had all been firmly established. But attempts to measure velocities and sizes of the particles had met with difficulties, even with the advent of the ultramicroscope, and it seemed that this method would not glean new proofs for the kinetic theory as a cause of Brownian movement. With this backlog of research to ponder, Perrin moved into a new direction for proofs of the kinetic theory of Brownian movement and of molecular reality, in fact, into a path totally independent from velocity determinations and from any involvement with questions of electrostatic charge. So successful was he in this fresh orientation that Duclaux, for example, while falling back on his own predispositions, praised Perrin for exacting an almost perfect harmony between experiment and the kinetic hypothesis.[16]

In his first detailed research on Brownian movement, Perrin used gamboge, which had been ever popular in nineteenth-century studies because of its bright colour and consequent ease of measurement. The lively animation of the Brownian movement was easily viewed under an ordinary microscope; after the emulsion was centrifuged, however, Brownian granules could only be observed with the lateral illumination of an ultramicroscope. Using a drop of the diluted emulsion in a microscopic preparation of 0·12 millimetres thick, the vertical distribution of the granules was noted after waiting several hours for equilibrium to be attained. A mean of several thousand readings resulted in the kind of distribution which Perrin had expected from his own sudden insight into Duclaux's casual observation that the distribution density of the granules was greater in the lower layers of a colloidal solution.[17]

For example, with a concentration of 100 for the granules at one level, he found that at levels 25, 50, 75, and 100 microns lower the density was represented by the numbers 116, 146, 170, and 200. If these latter numbers were 119, 142, 169, and 201, one would have a perfect geometric progression. Hence, Perrin concluded, the vertical distribution of the granules under conditions of equilibrium in this particular preparation, and probably in all colloidal solutions, is exponential, even as for a gas in equilibrium under the influence of its weight (aerostatic law of Laplace). In the atmosphere, the gaseous concentration is halved for an ascent of 6 kilometres; in a colloidal solution, it is halved for a height of 0·1 millimetre.[18]

In attempting to explain these experimental results, Perrin turned to the kinetic theory of gases, crediting Léon Gouy with the 'brilliant' hypothesis that Brownian movement is, in truth, a visible consequence of the molecular collision of the invisible liquid particles against the colloid particles. Imagine, postulated Perrin, a number of identical granules, where

$\rho =$ the density of the granules,
$m =$ the mass of the granules,
$n =$ the number of the granules per unit volume of water,
$k =$ the pressure of one granule.

Then, these granules exercise on any wall which arrests them (without hindering the water molecules) an osmotic pressure which is proportional to their concentration. Hence, if

$kn =$ the osmotic pressure due to the granules, and,
$ndh =$ the number of granules in one horizontal slice of height dh and unit cross-section,

then these ndh granules are kept in suspension by the sum of the force and the difference between the osmotic pressures on the two faces of the slice dh. Then

$$dn/n = (1/k)gdh(1 - 1/\rho)m$$

This integrates between the heights (0, h) to:

$$2.3\log(n_0/n) = (1/k)mgh(1 - 1/\rho)$$

showing that the distribution would, indeed, be exponential. Further, if one could find a value for m, one could calculate the constant k, the osmotic pressure of a single granule.[19]

In order to find m, Perrin utilised Stokes' law, applying it to a column of the emulsion in a vertical capillary tube, and observing the fact that when the emulsion is very far from equilibrium, the Brownian granules in the upper layers of the column fall as if they were droplets of a cloud. Using Stokes' formula relating the velocity of a spherical droplet, its radius, and the viscosity of the medium, Perrin found the radius of the granules.[20] From his own measurements of their density, he then derived a value of $k = 3{\cdot}60 \times 10^{-16}$. A few further calculations from kinetic theory quickly led him to the happy conclusion that the pressure exercised by a gas at the same temperature as the gamboge emulsion would be

$$n(RT/N) = n(3{\cdot}43 \times 10^{-16}),$$

and comparing the value of RT/N with that of his own k, Perrin concluded, 'Thus the suspended granules function like visible molecules of a perfect gas with a molecular weight around 3 milliards ($3{\cdot}3 \times 10^9$). The mean kinetic energy of a colloid granule is equal then to that of a molecule. . . . All at once the kinetic theory of fluids appears a little strengthened and molecules a little more tangible. Their number N per gram-molecule, deduced from the preceding equality, if it is assumed rigorous, is $6{\cdot}7 \times 10^{23}$.'[21]

The originality of Perrin's approach lay in his insight into a possible analogy between the behaviour of gas molecules and colloid particles which would lead to an irrefutable similarity in experimental phenomena and allow a clear derivation of fundamental values of kinetic-molecular theory. Certainly others, such as Léon Gouy, had anticipated that this could be done, but they groped in vain for a suitable method and often became bogged down in futile attempts to determine the actual velocities of the Brownian granules. Not only was Perrin's conceptual approach of 1908 entirely distinctive, but his experimental techniques were far superior to those who had laboured before him. In a letter to Perrin praising the 1909 memoir 'Le mouvement Brownien et le réalité moléculaire', Gouy acknowledged Perrin's ingenious procedure of preparing uniform granules of gamboge, confessing 'I can tell you myself that, if it had not been for the difficulty which I perceived in preparing such granules, I would have more actively pursued my experimentation on Brownian movement.'[22]

The technical execution of the experiments was then extremely important to the proofs of his point, since a high degree of accuracy was necessary for a large body of data. The four papers published in 1908, all dealing exclusively with the height distribution, were entirely the work of Perrin, as it was not until a fifth series of experiments that he enlisted the aid of his graduate students.[23] After the first paper in this group, Perrin was accused by Duclaux of an overweening simplicity in expressing the emulsion concentration as a function of height (dn/dh) when each particle was supposedly regulated by the laws of probability: '... it is to be feared that the simplicity of reasoning ... harms its accuracy'.[24] In such a context, one recalls the comment by Paul Painlevé that 'simplicity can be obtained only at the cost of precision. But I do not believe that one can be more precise than Monsieur Perrin, if one also wants to remain simple.'[25]

Such precision Perrin did indeed attain in his first experiments on Brownian movement, and he continued to perfect his experimental technique in the next year as he achieved ever more accurate results. Not the least of his problems in such a feat was the preparation of a satisfactory emulsion containing Brownian particles of identical diameters. His first forays against this challenge were undertaken with ordinary colloidal solutions of arsenic sulphide and ferric hydroxide.[26] When successive failures accompanied this approach, he turned to the dried vegetable latex gamboge. When a stick of gamboge was rubbed by hand into some water, it formed a bright yellow emulsion of glass-like spherical grains, which he then treated with methyl alcohol, dissolving $\frac{4}{5}$ the weight of the material. After centrifuging the mixture to separate the alcohol and gamboge, he diluted the latter with pure water, forming an emulsion with granules of less than 1 micron diameter. In later experiments he sometimes used emulsions of mastic, which were handled in much the same fashion.[27]

The process of centrifugation was pivotal to the validity of the entire investigation since it was a prime necessity that the granules be of equal diameter when their parameters were substituted into the exponential function for height distribution. Perrin's procedure was analogous to fractional distillation; he first filled a test-tube to a level H with the emulsion. Using a table to predict minute by minute, for each temperature and each velocity of rotation of the centrifuge, the distance which a grain of given radius would fall, he summed these successive distances equivalent to the original height H. After centrifuging, a first sediment S_1 remained after the first liquid L_1 was decanted. L_1 was then distributed into several test-tubes, while the sediment S_1 was diluted with water to the height H. With successive centrifugations, decantations and dilutions, all the grains with radius between $(a–e)$ and (a) were gathered in the sediment S_2, all smaller grains in the portions L_2, the larger grains in S_1; the sediment S_2 therefore contained an emulsion practically uniform. Such a process required a period of several months in which 1200 grammes of gamboge eventually produced several decigrammes of uniform granules.[28]

A second problem, once the emulsion had been prepared, lay in ascertaining the density of the granules (ρ). Indeed Perrin's density figures in his first paper were slightly in error, as recognised by Duclaux, jeopardising the accuracy of the value of k.[29] As described in a later paper, Perrin perfected two principal methods for density determinations. In the first, he dried the gamboge at 130° C and then quickly cooled it to produce a transparent glass. The fragments of this glass were placed in water and potassium bromide added until the fragments became suspended, indicating that the density of the glass bits was the same as that of the solution. In a second method, Perrin measured the masses m of the water and m' of the emulsion, using the same specific gravity flask. The mass μ of the gamboge resin in the mass m' was determined by drying, so that $(m' - \mu)$, the mass of the intergranular water, might be calculated. Since the volume of the flask was m/d, that of the intergranular water $(m' - \mu)/d$, and that of the granules $(m/d - [m' - \mu]/d)$, then $\mu/(m/d - [m' - \mu]/d)$ gave the required density of the granules.[30]

In assembling the emulsion for observation, a droplet was placed in a Zeiss cell and covered by an object-cover which closed it completely. The preparation might then be placed vertically and the body of the microscope horizontally, or the preparation horizontally and the body of the microscope vertically. In the first series of experiments Perrin used an ultramicroscope, but in the next nine series, he most often worked with an ordinary microscope.[31] The objective was adjusted to a small depth of focus in order to view clearly granules in a horizontal layer one micron in depth; alternate layers could be seen by raising and lowering the microscope. The value h was determined by reading the displacement h' from a graduated disc fixed to the micrometer screw on the microscope, and a correction factor was applied involving the refractive index of the two media separated by the cover glass.[32]

Certainly the trickiest and most time-consuming dilemma was the actual counting of the particles in the preparation and the evaluation of n_0/n. To achieve this end, Perrin employed a needlepoint field, where the observer viewed the preparation through an opaque disc of tin foil pierced by a dissecting needle. In this very limited field, one could see at one instant an exact number of grains, usually less than five or six. With the use of a timed camera shutter, instantaneous photographs might be taken over this very small range, noting successively the number of grains in one layer, for example,

3, 2, 0, 3, 2, 2, 5, 3, 1, 2.

Then the observation was repeated at another level, producing such figures as

2, 1, 0, 0, 1, 1, 3, 1, 0, 0.

Due to the irregularity of the Brownian movement, 200 readings of this kind might replace an instantaneous photograph covering a field 200 times more extensive. After averaging the number of granules for a given volume and time interval, the comparison of the frequencies at two levels resulted in n_0/n, the desired ratio of concentrations.[33]

With granules of diameter larger than 0·5 micron, it was possible to use instantaneous photographs of the entire field and average the number of sharp images. Perrin first successfully employed this method for an article published in 1909.[34] The visual results of this procedure, obtained by placing above one another five photographic sections made at intervals of 6 microns in an emulsion of gamboge, were quite convincing, if not breathtaking, in their portrayal of the validity of the exponential law as derived from kinetic theory (see Fig. 3, p.108).[35]

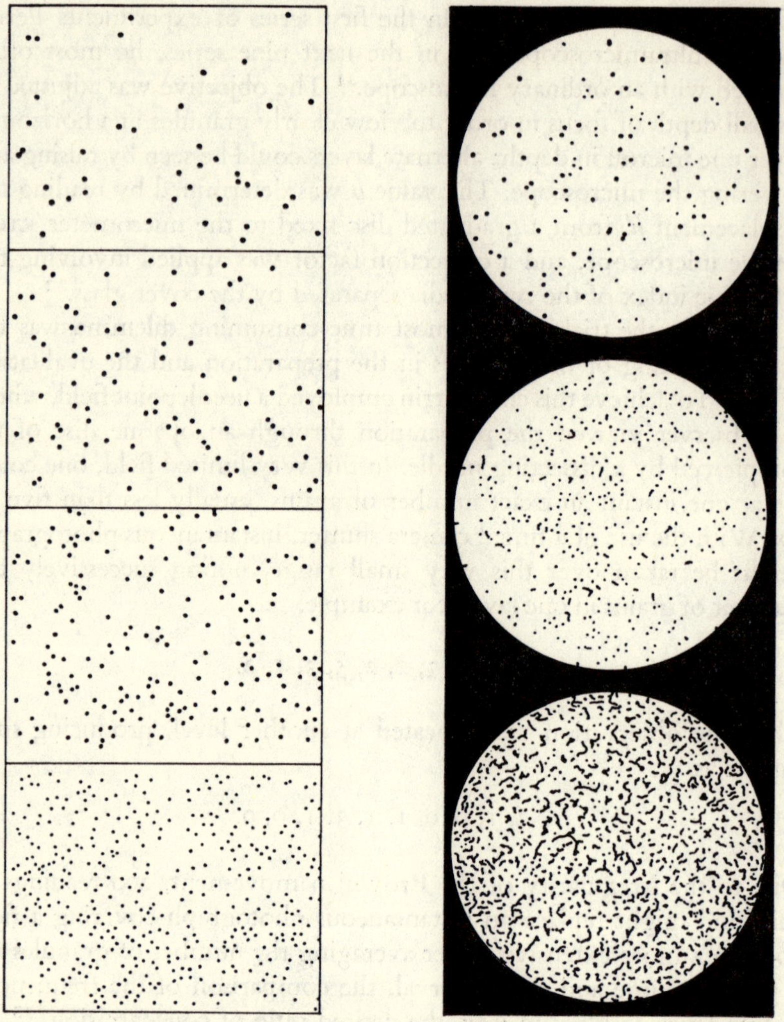

FIG. 3

Perhaps the most cogent criticism of Perrin's first paper was directed at the next step in his procedure, the determination of the radius *a* of the granules. Duclaux was apprehensive that Perrin's observation of free fall and application of Stokes' formula were terribly hazardous for particles of such small dimensions,[36] and his critique goaded Perrin into double-checking the method's appropriateness. After recalling that both J. J. Thomson and Paul Langevin had applied Stokes' equation to particles of the order of size of the Brownian granules—admittedly conditions far from the case of continuity originally allowed by Stokes—Perrin detailed in his second paper on Brownian movement a control experiment for his previous determination of the radius of the particles.[37]

Having observed that the emulsion granules adhered to the glass walls of the object- and cover-glasses in a slightly acidic solution, Perrin perceived that the number of granules in a preparation of known height (about 100 microns) and known density (discerned from a prior titration of the emulsion) could actually be counted, giving the individual radius. If the particles were large enough, one might even measure the mean length of a file of 5–6 granules and more easily determine the radius. Carrying out these measurements, Perrin concluded that the law of Stokes was valid for radii of the order of 0·1 micron and probably well below that order.

In additional experience with the emulsion particles, Perrin also discovered a third method for computing the value of radius *a*, dependent upon the evaporation of a very dilute drop of emulsion upon the object-glass. By capillary action, the grains assembled in a single thickness, like bullets in a horizontal slice of a range of bullets. Counting the number of grains aligned in a field of measured length or covering a regular surface provided a rough measurement of the diameter.[38]

Perrin's third paper on Brownian movement likewise answered certain criticisms of Duclaux and made another important analogy between the behaviour of emulsions and gases under the limits of the kinetic theory. (This paper also included the affirmative results of a new series of observations on the height distribution, using granules twice as large as in the first series, i.e., 0·21 micron instead of 0·12 micron.) In reply to the assertion that Perrin's values of osmotic pressure differed from those in much of the colloid literature, Perrin differentiated between dilute and concentrated emulsions. Whereas most measurements of osmotic pressure had been applied to concentrated emulsions, Perrin's were pursued with only dilute emulsions, where the volume of the granules was less than $\frac{1}{4}$ the intergranular volume. The osmotic pressure due to the Brownian movement would cease to be proportional to their concentration for a more concentrated emulsion. Drawing the parallel to gas laws, Perrin chided that his calculations would no more apply to concentrated conditions than would Mariotte's (Boyle's) law to air at the density of water.[39]

By the time of the appearance of Perrin's fourth 1908 article on

Brownian movement, he could cite three series of experiments, involving calculations for 13 000 granules—3000–4000 granules per series—in 16 000 readings.[40] With such factual material firmly under his belt, he felt licensed to extrapolate his results into calculations of what he termed the 'constantes universelle'. He explained:

> The number of molecules N that the total gram-molecule contains, the charge e of the electron and the quotient α of the mean energy of a molecule by its absolute temperature T are the *universal constants* which are all known as soon as any one of them is known.[41]

Previous calculations of these constants had been effected from various approaches. From the limits imposed upon molecular volumes by assumptions about the packing of bullets and the packing of conducting spheres, the value of N had been fixed between 4×10^{22} and 204×10^{22}. Such arguments supposed that molecules in a liquid could never be more tightly packed than the bullets in a pile of bullets, and that the true volume of the molecules could be no larger than that of perfectly conducting spheres which, substituted for these molecules, gave the same dielectric constant to a fluid. From the same school of classical kinetic theory, Van der Waals reasoned that the true volume of N molecules must be $\frac{1}{12}$ the critical volume, making N approximate 60×10^{22}.

More carefully bounding the value of N, Perrin wrote, were the recent estimates from the Cavendish laboratory's use of the cloud chamber, in which water vapour was condensed on to gaseous ions and the charge determined from observations of the charged droplets in an electric field. These investigations placed N between 43 and 96×10^{22}. The newest experiments of Rutherford on radioactivity implied limits of 62 and 73×10^{22}, while a third line of research by Max Planck and H. A. Lorentz on the theory of black-body radiation had pinned down values of $N = 61 \times 10^{22}$ and $N = 77 \times 10^{22}$, respectively.

A superior precision, Perrin affirmed, existed in the values from his studies of Brownian movement. Here, through a series of three experiments—'directe et susceptible d'une précision illimitée'[42]—was obtained a mean value for N of 71×10^{22}, implying values for e of $4{\cdot}1 \times 10^{-10}$ e.s.u. and for α of $1{\cdot}7 \times 10^{-16}$. The mass of a molecule or ordinary atom might also be figured with the same exactness, as well as molecular diameter and other values. Through the experiments on Brownian movement, accurate values for the fundamental parameters of the kinetic-molecular theory could be computed, and their general agreement with values derived from totally different experimental and theoretical approaches would incontrovertibly confirm the reliability and the truth of the molecular picture in general.

The Physicists' Theories of Displacement and Rotation

Perrin had now successfully answered the questions first broached in his paper of 1905, regarding the osmotic pressure and the mean kinetic energy of colloid particles, as well as their parallel behaviour to the particles of a gas or a liquid at equilibrium. He had also developed a mathematical explanation of the height distribution first alluded to in 1905, and he had carried out accurate measurements of this physical distribution which enabled him to calculate a precise mean value for N and to extrapolate it to accurate values for e and α, realising in great part his ambitions to prove the validity of the kinetic theory and of molecular reality. Perrin also soon comprehended that his experimental work established empirically the hypothesis of the equipartition of kinetic energy used by Einstein, and as Perrin studied Einstein's work more carefully, he grasped that he had verified a relationship in Einstein's theoretical derivations that actually predicted the very exponential height distribution for large particles which he had now so convincingly observed.

In his first paper on Brownian movement, Perrin alluded to the equipartition hypotheses of Einstein and Langevin, although he developed the general remark no further. Both Einstein and Smoluchowski, in fact, had recognised by 1906 that an exponential distribution of Brownian particles was a necessary consequence of the equipartition of energy, but they apparently failed to visualise any *experimentum crucis*. In turn, Perrin remained unaware of their predictions until his early experiments had already been accomplished.[43] Langevin had certainly directed Perrin's attention to Einstein's papers by 1908, but Perrin did not at first study them carefully enough to realise that Einstein had himself presaged an exponential height distribution.[44]

That this was true is certainly not surprising when one closely surveys Einstein's papers of 1905 and 1906, for their mathematical derivations were certainly beyond the ken of even the more precocious experimentalist; Einstein's methods were novel as well as ingenious. Certainly the fundamental displacement relationship for Brownian movement which was evolved by Einstein was known by most physicists and chemists studying colloidal suspensions; for example, Cotton and Mouton included in their text of 1906, *Les Ultramicroscopes et les Objets Ultramicroscopiques*, the Einstein equations relating particle displacement during a given interval to the square root of the time and to the inverse root of the viscosity.[45] But Perrin, on the basis of experiments by Victor Henri and others, was suspicious of these equations' exactness, questioning the appropriateness of Stokes' law in its application by Einstein.[46] In the first publication emerging from Perrin's laboratory and written by his student Chaudesaigues as part of the latter's doctoral prerequisites, Chaudesaigues remarked that Perrin's 1908 experiments and calculations had produced a precise measure of *N* and had confirmed Gouy's hypothesis, but had not affirmed the complementary hypothesis of Einstein. Chaudesaigues worked, of course, under Perrin's direction in this matter, and it was Perrin who prepared and measured the diameters of the spherical granules of gamboge employed by his student.

At the time of Perrin's investigations of Brownian movement, there were three principal theoretical treatments of the problem—those of Einstein (1905, 1906), Smoluchowski (1906) and Langevin (1908).[47] Smoluchowski investigated the life history of one Brownian particle in irregular collisions with the suspending fluid's particles (using time averages), while Langevin studied the life history of not just one particle, but of a whole ensemble. Einstein, on the other hand, dealt with a probability distribution from which time averages could be calculated.[48] Einstein's derivation of the displacements of Brownian movements was in a sense the least generalised and abstract, in that his first 1905 paper assailed the problem from considerations of osmotic pressure, diffusion and the variation in the distribution density of suspended particles under the influence of gravity.

His first paper, 'Über die von der molekular-kinetischen Theorie der Wärme geforderte Bewegungen von in ruhenden Flüssigkeiten suspendierten Teilchen',[49] was not ostensibly directed toward an explanation of Brownian movement *per se*, though he realised that the problem he was dealing with was most probably identical to the Brownian movement. In the absence of any exact treatises on the subject, however, he refused to make such a judgment and prefaced his theoretical work by stating that

> In this paper it will be shown that according to the molecular-kinetic theory of heat, bodies of microscopically-visible size suspended in a

> liquid will perform movements of such magnitude that they can be easily observed in a microscope, on account of the molecular motions of heat. It is possible that the movements to be discussed here are identical with the so-called 'Brownian molecular motion'.

Einstein continued:

> If the movement discussed here can actually be observed (together with the laws relating to it that one would expect to find), then classical thermodynamics can no longer be looked upon as applicable with precision to bodies even of dimensions distinguishable in a microscope; an exact determination of actual atomic dimensions is then possible. On the other hand, had the prediction of this movement proved to be incorrect, a weighty argument would be provided against the molecular-kinetic conception of heat.[50]

So Einstein himself posed the problem which Perrin independently promulgated, and Einstein, as well as Perrin, fully appreciated the extraordinary effect which an experimental proof relating a phenomenon such as Brownian movement to the kinetic-molecular theory would have in securing the complete acceptance of that theory and in dumbfounding its critics.

Indeed, the energeticists and other opponents of the discontinuous hypothesis had much to learn from studies of colloidal solutions.[51] As Einstein remarked, the validity of classical interpretations of thermodynamics for particles was wholly at stake, for that viewpoint, as he understood it, did not predict an osmotic pressure from large particles suspended in a liquid. Einstein wrote that if, for example, one considers a partial volume V^* of a liquid of total volume V, in which are dissolved z gram-molecules of a non-electrolyte, and one immerses in the solution a semipermeable membrane through which the dissolving medium—but not the dissolved substance—can move, then an osmotic pressure is predicted by kinetic interpretations, such that

$$pV^* = RTz$$

But, if instead of the dissolved substance, small suspended particles are in the volume V^*, which likewise cannot penetrate the membrane, then one does not expect a force to be exerted on the membrane, for the free energy—granting the traditional conception of thermodynamics—does not depend upon the position of the membrane and suspended particles, but only upon the total mass and properties of the suspended substance, the liquid and membrane, as well as upon the temperature and pressure.[52]

From the standpoint of the molecular-kinetic heat theory, however, one comes to another conception, as Einstein points out. According to this theory, dissolved molecules differ from suspended particles only by their size, and one cannot comprehend why a number of suspended particles should not exert the same osmotic pressure as the same number of dissolved molecules. Considering, then, a system of variables of state $p_1, p_2, \ldots, p_n$, Einstein computes the entropy and the free energy of the system. He then asks how the free energy of the system will be affected by the pressure of colloidal particles represented by $V^\star$.

Given $F = (RT/-N)\ln B$, where $B = \int \exp(-EN/RT)\, dp_1 \ldots dp_n$,

and $dB = (J dx_1 dy_1 \ldots dz_n)$,

Einstein demonstrates that J is independent of the centres of gravity of the particles $(dx_1, dy_1 \ldots, dz_n)$ and of the position of the semipermeable membrane.

$$\text{Since } B = \int J dx_1 \ldots dz_n = JV^{\star n}$$

$$\text{then, } F = (-RT/N)\{\ln J + n \ln V^\star\}$$

The free energy F is therefore dependent upon the position of the membrane, and the osmotic pressure p is given by

$$p = (-\partial F/\partial V^\star) = (RT/V^\star)n/N$$

Hence, an osmotic pressure may be derived from the molecular-kinetic theory of heat, and one must conclude that theoretically, solute particles and suspended particles act identically at great dilutions.[53]

In addition, Einstein analyses the dynamic equilibrium of large particles suspended in a fluid as the superposition of two processes operating in opposite directions: the movement of the suspended substance under the influence of a force K [gravity] acting on each suspended particle, and a process of diffusion, viewed as a result of the irregular movement of the particles produced by the thermal molecular movements. Hence, (from Stokes' law) the force K imparts to spherical particles of radius P in a liquid of viscosity k a velocity $K/(6\pi kP)$, so that $(\nu K/6\pi kP)$ particles pass by a unit area in unit time (where $\nu = n/V^\star$).

Where D is the coefficient of diffusion of the suspended particle and μ is the particle mass, there crosses a unit area in unit time—due to diffusion—a total of $-D\partial(\mu\nu)/\partial x$ grammes or $-D\partial\nu/\partial x$ particles. Since there must be a dynamic equilibrium of these two processes,

$$\frac{\nu K}{6\pi kP} - \frac{D\,\partial\nu}{\partial x} = 0 \qquad (1)$$

From equation (1) and equation (2) (which expresses the over-all criterion of equilibrium that $dF = dE - TdS = 0$),

i.e. $$-K\nu + \frac{RT}{N}\cdot\frac{\partial\nu}{\partial x} = 0 \qquad (2)$$

it follows that: $$D = \frac{RT}{N}\frac{1}{6\pi kP} \qquad (3)$$

This gives the diffusion coefficient for the condition that a molecule of the dissolved substance is large compared with a molecule of the solvent; no quantities depending on the nature of the solution appear except the viscosity of the solvent and the diameter of the solute molecules.[54]

Einstein also relates the process of diffusion to the irregular motions of the solute molecules by utilising a Boltzmannian statistical interpretation. Assuming that the number of particles dn which undergo a change Δ in their x-coordinates in the time interval τ adhere to the probability law that

$$dn = n\,.\,\phi(\Delta)d\Delta$$

where $\int_{-\infty}^{+\infty}\phi(\Delta)d\Delta = 1$ and $\phi(\Delta) = \phi(-\Delta)$, he calculates the distribution for n particles after a time interval τ. This gives, for $f(x,t)$:

$$\frac{\partial f}{\partial t} = D\frac{\partial^2 f}{\partial x^2}$$

the familiar equation for diffusion.[55] Since the assumption of an irregular motion of large particles leads to this diffusion relationship, one may study the function $f(x,t)$ from the standpoint of diffusion outwards from a point, utilising the relation

$$f(x,t) = \frac{n}{\sqrt{4\pi D}}\frac{(\exp -x^2)/4Dt}{\sqrt{t}}$$

to find the average displacement of a particle in the x-direction. Since

$$n = \int_{-\infty}^{+\infty} f(x,t)dx$$

then
$$x^2 = 1/n \int_{-\infty}^{+\infty} f(x,t)x^2 dx$$

and
$$x^2 = 1/(4\pi Dt) \int_{-\infty}^{+\infty} (\exp -x^2/4Dt)x^2 dx$$

Hence,
$$x^2 = 2Dt$$

or
$$\lambda_x = \sqrt{x^2} = \sqrt{2Dt} \tag{4}$$

Through such a derivation[56] Einstein defines 'the displacement λ_x [rather than velocity] in the direction of the x-axis which a particle experiences on the average, or—more accurately expressed—the square root of the arithmetic mean of the squares of displacements in the direction of the x-axis. . . . The mean displacement is therefore proportional to the square root of the time.'[57]

Having related the average magnitude of the motions of the particles to the diffusion coefficient (in (4)), it is then possible to calculate the displacement in terms of the viscosity of the solvent and the size of the solute molecules (from (3)), with the results applying both to actual dissolved molecules and to any particles suspended in a liquid. From the formulas for the diffusion D (3) and the displacement λ_x (4), one may deduce:

$$\lambda_x = \sqrt{t}\sqrt{\left(\frac{RT}{N} \cdot \frac{1}{3\pi kP}\right)} \tag{5}$$

the equation which, along with that for λ_x in terms of the diffusion coefficient (4), was later to be most frequently quoted from this paper.

After suggesting the order of magnitude which such an equation predicts (for example, $\lambda_x = 0{\cdot}8$ micron for $t = 1$ second, $N = 6 \times 10^{22}$, $T = 17°$ C, $k = 1{\cdot}35 \times 10^{-2}$, and $P = 0{\cdot}001$ millimetres), and after proposing that an exact evaluation of N could also be obtained by observations of λ_x, Einstein concluded his historic paper of 1905 by hoping 'that some enquirer [sic] may succeed shortly in solving the problem suggested here, which is so important in connection with the theory of heat'.[58]

Subsequent papers in 1906 and 1907 dealt further with the general conceptions of this first paper on suspended particles. In his second paper on Brownian movement (*Annalen der Physik*, 1906), Einstein attempted to determine the size of the solute molecules in an undissociated dilute solution from the viscosities of solution and pure solvent, and from the rate of diffusion of solute into solvent, assuming the volume of the solute molecules to be large compared with the volume of the solvent molecules. In carrying out this procedure, he abstracted values for the diffusion coefficient and the viscosity in a one per cent aqueous sugar solution from the tables of Landolt and Börnstein, obtaining a mathematical result of $N = 4{\cdot}15 \times 10^{23}$.[59]

By the time of Einstein's second publication of 1906, he had been informed by Siedentopf that earlier observations of Brownian movement by both Siedentopf and Gouy confirmed the order of magnitude of Einstein's calculations. Einstein had also noted the rumbling criticisms that his method of derivation was too specific and restricted. He now endeavoured to present a more general method to show how Brownian movement is related to the fundamentals of the molecular theory of heat and to develop formulas for both translational and rotational movement. Therefore in his third paper on Brownian movement (also in *Annalen der Physik*, 1906), his general format was quite removed from physical realities, with osmotic pressure and diffusion left unmentioned. Rather, he spoke only of probability, energy, and irregular thermal processes, and instead of signifying a diffusion coefficient D, he used the letter B—'the mobility of the system in respect to the parameter α'—('die Beweglichkeit des Systems in bezug auf den Parameter α').[60]

It was in this third paper that Einstein derived the mean rotation (α) of a suspended spherical granule as the result of irregular thermal processes, arriving once more at an equation of displacement (not of angular velocity):[61]

$$\sqrt{\bar{\Delta}_r^2} = \sqrt{t}\sqrt{\left(\frac{RT}{N}\cdot\frac{1}{4\pi k P^3}\right)} \qquad (6)$$

In analysing the question of how far a parameter of a system can diverge from the value corresponding to stable equilibrium when a constant external force is applied—a problem formulated as an exponential law by Boltzmann—Einstein arrived at an exponential relation which Perrin later recognised as the theoretical derivation of the phenomenon which he had experimentally observed. What Einstein did was to apply Boltzmann's expression to determine how small a suspended particle must be in order to remain permanently suspended despite the pull of gravitation. Where the particle is of greater density than the liquid (the opposite case being fully analogous), he found that with dW the probability, the parameter α lies between α and $(\alpha - d\alpha)$, and

$$dW = \text{const.}\{\exp\left[\frac{-N}{RT}v(\rho - \rho_0)gdx\right]\}dx$$

where v = the volume of the particle, ρ = its density, ρ_0 = the density of the liquid, and x = the vertical distance of the particle from the bottom of the vessel.[62]

Besides these three papers on Brownian movement, Einstein published two additional ones in 1907 and 1908, though neither contributed anything substantially new to theory. The 1908 paper met a widespread scientific demand for a simplified version of the theory,[63] while the paper of 1907 refuted the assumption of The Svedberg that one could actually observe the velocity of a Brownian particle. In this connection, Einstein computed that the time for the velocity of a particle of colloidal platinum, for example, to fall to $\frac{1}{10}$ of its original value would be $3 \cdot 3 \times 10^{-7}$ seconds. Yet, Einstein reasoned, one must assume that the particle receives new impulsions to movement by some process inverse to viscosity, so that the particle retains an average mean square velocity; the direction and magnitude of these impulsions are independent of the original direction of the particle's motion and velocity, however. The velocity and direction of motion of the particle will then already be very greatly altered in an extraordinarily short time (and in a totally irregular manner), and it is therefore impossible—at least for ultramicroscopic particles—to ascertain v^2 by observation.[64]

Einstein's 1905 publication was promptly succeeded by that of Maryan Smoluchowski, who was to become director of the Institute of Physics at the University of Cracow in 1913.[65] His derivation of an equation for the displacement of a particle large in size in comparison with that of liquid molecules proceeded from an analysis of the life history of a single particle, as mentioned earlier. His method resulted in the equation:[66]

$$\Delta = \frac{8}{9\sqrt{\pi}} \cdot \frac{c\sqrt{m}}{R\sqrt{\mu}}$$

(Δ = the displacement of the particle in the x-axis direction, m = the mass of the liquid molecule, c = a gas theory constant, R = the radius of the suspended particle, and μ = the viscosity of the liquid), which was in essence almost identical to the equation of Einstein, though with a coefficient $\frac{27}{64}$ smaller.[67] Unlike Einstein, Smoluchowski was well acquainted with the earlier qualitative literature on the phenomena of Brownian movement and included a detailed bibliography of such studies in his paper.

Since he himself had not attempted experimental verification of the equation, Smoluchowski sought partly to remedy this deficiency with a presentation of the older studies, which he felt demonstrated that the Brownian phenomenon was identical with theoretically conceived *Molekularbewegungen.* Most of the previous experimental work, he believed, had been largely negative, merely excluding certain explanations of Brownian movement as implausible. For example, the phenomenon's independence from evaporation, from convection currents and light had been elucidated. If electrical forces were the causal ones, Smoluchowski further reasoned, there should be a grouping of the granules rather than an irregular motion; capillary energy, too, was an unacceptable explanation, especially since it could in no way account for the Brownian movement of microscopic bubbles in mineral cavities. The appropriateness of the kinetic theory, on the other hand, Smoluchowski indicated, was in keeping with the observed ratio of the amplitude of motion to particle size, the effect of viscosity on the movement, and its invariability with time.

According to Smoluchowski, the only sensible objection to the efficacy of the kinetic theory in explaining Brownian movement had been Nägeli's protest that velocities due to molecular collision would be too small to be observed. In countering this remonstrance, Smoluchowski proposed an analogy that was to become quite popular in later discussions of Brownian movement. That this motion could not be seen, proffered Smoluchowski, was as mistaken as believing that a dice-player can never lose a greater sum than the stake set out for a single throw. If one says that there is an equal probability of winning and losing, and that for n throws, m are successful and $(n - m)$ unsuccessful, then one has $n!/\{2^n m!(n - m!)\}$ as the probability of breaking even.

The mean deviation, positive or negative, from the null value amounts to:

$$v = 2\sum_{m=n/2}^{n} \frac{2m - n}{2^n} = \frac{n}{2^n}\left|\begin{matrix} n \\ n/2 \end{matrix}\right.$$

which for large values of n, becomes,

$$v = \sqrt{\frac{2n}{\pi}},$$

demonstrating that one's gain (or loss) tends to become proportional to the square root of the number of hands played.

Similarly, the velocity $C = c\sqrt{m/M}$ imparted through a collision to a particle of mass M by a molecule of mass m will be very small, and the mean in the x-direction is merely a fraction of that. However, the sphere M suffers such collisions 10^{16} times per second in air and 10^{20} times per second in water, and although there is some compensation, to be sure, in relation to the momentum in the x-direction—indicating a positive or negative surplus of 10^8 or 10^{10}—one must conclude that every particle M preserves a velocity of 10^2 or 10^4 cm/s in the direction of $+x$ or $-x$. That is, probabilistic considerations from the kinetic theory negate Nägeli's critique.[68]

Like Einstein, Smoluchowski felt compelled to cope with the older values of velocity, which were in disagreement with the new formulas and with Maxwellian velocities.[69] As Einstein, he explained that the particles could not possibly be followed directly, but that only a mean position of the particles could be observed. The centre of gravity of a granule would actually describe an irregular, complicated zig-zag path whose rectilinear parts were smaller than the dimensions of the particle itself. Only in so far as the geometrical sum of these lines might result in a certain noticeable size through a period of time would a displacement be noticed.[70]

Smoluchowski concluded his paper with a calculation of the diffusion coefficient and the osmotic pressure which had been the starting point for Einstein in 1905. From considerations of the stability of a suspension, he, too, arrived at an atmospheric pressure formula: $N = N_0 \exp - d^2$, with

$$a = \frac{4\pi R^3}{3} \cdot \frac{(\sigma - \sigma)g}{\mu r_0 \theta}$$

This formulation, he remarks, 'suggests that a lasting stability which one can call "real" would be noticeable with microscopically visible particles only in exceptional cases on account of the size of the coefficient, and it would play a more important role primarily with many smaller particles (for example, for metal colloids of radius 10^{-6} cm)'.[71]

Initiated after seeing Einstein's 1905 publication, Smoluchowski's paper includes the claim that his results had been accomplished several years previously. In addition, he considered his method more direct, simpler and perhaps more convincing than Einstein's.[72] He realised that, as with Einstein's derivation, his use of Stokes' formula could be questioned since it rested on the assumptions that the capillary forces were null and that the particles were absolutely rigid spheres. Smoluchowski was also later criticised for his incorrect interpretations of the equipartition theorem as well as for certain mathematical errors in his 1906 paper (which had fortunately cancelled each other out).[73] There is little doubt, however, that his 1906 paper and later publications were influential in deciding the controversy between statistical and traditional concepts of thermodynamics and that he achieved his aim of providing important arguments for the molecular-kinetic nature of matter.[74]

The third major paper of the decade on the theory of Brownian movement was published by Langevin in the *Comptes rendus* in 1908, shortly before Perrin's first experimental paper.[75] Langevin was aware of both Einstein's and Smoluchowski's approaches, and intended to establish that a correct application of the method of Smoluchowski would lead to the exact formulation of Einstein—with the same coefficient—and to demonstrate that an altogether different method unravelled an infinitely more simple demonstration. As with his predecessors, Langevin's point of departure was the equipartition of kinetic energy among various degrees of freedom in a system of thermal equilibrium. He split the forces exercised upon the suspended particles into two parts, those due to gravity and those due to molecular impact, eventually leading to Einstein's equation in the form,

$$\overline{\Delta}_x{}^2 = \frac{RT}{N} \cdot \frac{1}{3\pi\mu a\tau}$$

This derivation was achieved through considering a large number of identical particles and taking the mean of equations written for each of them, eliminating one term because of the irregularity of motions and another because of its minuteness. Langevin's complete derivation, though not detailed here, is the one normally presented in textbooks, although it, too, was criticised at the time of its inception.[76]

Doubt Creeps In: A Failure in Verification

No sooner were the theoretical papers of Einstein and Smoluchowski published than endeavours linking experimental observations of Brownian movement with the theoretical laws appeared. Unfortunately for the theory, the most cohesive aspect of these various endeavours was the lack of anything resembling a clear verification. Of those who attempted a conjunction of the theoretical and experimental aspects of Brownian movement, certainly the most prolific was Svedberg, who presented his inaugural dissertation at the University of Uppsala based on investigations of colloids and the *Molekularbewegung*, which he had observed in the previous several years. In Germany, in the laboratory of Franz Richarz, Max Seddig performed some researches on the phenomena, and in France, Victor Henri unsuccessfully attempted cinematographic studies to confirm Einstein's mathematical description.[77]

Certainly Svedberg believed that his efforts provided a direct and affirmative response to the pleas of Einstein, Zsigmondy and Siedentopf for accurate data on the magnitudes of Brownian movements.[78] Svedberg, like Perrin, was originally motivated to provide arguments bolstering the molecular-kinetic conception, and this purpose infused his earliest observations.[79] His researches began in 1905, when he initiated studies of the colloidal distribution of materials of constant particle size in various solution media. His first major paper on the subject appeared in the *Zeitschrift für Elektrochemie* describing what he called the 'amplitude' of the movements as well as the mean absolute velocity.[80] It was this publication which Einstein criticised for the calculation of a velocity which he himself felt to be totally unobservable. In the second part of Svedberg's 1906 paper, the experimenter introduced two general formulations which he had first postulated before becoming acquainted with Einstein's 1905 paper. These relations, as well as his actual experimental data, Svedberg attempted to relate to Einstein's equations.

In viewing various hydrosols and organosols with a vertical ultramicroscopic apparatus—a system in which the fluid slowly dripped downwards at constant velocity through a capillary tube—Svedberg saw a motion of the particles which seemed to resemble a sinusoidal light curve ('Beobachtet man unter Anwendung dieser Vorrichtung eine kolloide Lösung, so erscheint das Gesichtsfeld von unzähligen sinusoïdähnlichen Lichtkurven durchsetzt').[81] In justifying such visual curves of well-defined amplitude and wavelength, Svedberg relates the following explanation: that one may decompose the zig-zag movements of the suspended particles into directions parallel and vertical to the movement of translation. The velocity component parallel to an initial direction of translation then increases or decreases the velocity of this translation; certain vertical components also exert an effect. If the relations of movement are sufficiently stationary during the time required for a set of particles to pass the field of view of the microscope, a well-defined path results, under the influence of the distribution velocity of translation. Because of the durability of light impressions on the retina of the eye, this path is observed as an unbroken whole by the observer.[82]

Studying emulsions of platinum in aqueous and organic solvents, Svedberg estimated amplitudes (A) of the sinusoidal path through comparison with the microscope's ocular scale. He then plotted the quantity $2A$ for eight different emulsions against increasing viscosities of the solvents, arriving at a hyperbolic curve. Determining certain parameters for the curves and observing the interrelationship of these values led him to the enunciation of two laws: the statement that $A \, . \, n = k$ and $a/\tau = k'$;[83] where $A =$ the amplitude of the sinusoidal curve, $\tau =$ the period of a particle's [Brownian] movement as it is displaced a distance $2A$, $n =$ the viscosity of the solvent and k and k' are two different constants.*

*One way of understanding these parameters is to visualise a curve representing the over-all particle motion viewed by Svedberg, a motion due both to the regular dripping of the emulsion through a capillary tube connected to the vertical glass cuvette and due also to the irregular Brownian movement of the emulsion granules. Consider at first three principal parameters: $\lambda =$ the wavelength of the curve, in cm; $A =$ the amplitude of the curve, in cm (but both are of the order of microns); and $H =$ what Svedberg terms the 'Translationgeschwindigkeit', in cm/s, i.e. the total velocity of movement of the emulsion as it drips through the capillary tube, with Brownian movement contributing to this total velocity. Svedberg then defines $\tau = \lambda/H$ as the period ('Schwingungszeit') (in seconds), apparently a measure of the time spent by the particles in irregular movement independent of the fact that the total emulsion is moving (dripping) downwards. According to Svedberg, the value $4A/\tau$, in cm/s, is then a measure of the absolute mean velocity of the suspended particles, due to their irregular Brownian movement.

After reading Einstein's 1905 paper, Svedberg attempted to relate Einstein's equation for displacements of translation to his own laws, assuming $\Delta_x = 4A$, $t = \tau$, and $k = n$. If one should then write Einstein's equation as

$$4A = \sqrt{\tau}.\sqrt{\left(\frac{RT}{N}\cdot\frac{1}{3\pi np}\right)}$$

and gather the constants into one expression, one would have:

$$A = C_1\sqrt{\frac{\tau}{n}}, \text{ and } A/\tau = C_2\,\frac{1}{A\,.\,n}$$

Recalling that (Svedberg's) experimental data indicates A/τ as constant, then

$$C_2\cdot\frac{1}{An} = \text{const., and } A\,.\,n = \text{const.}$$

The latter is the equation of a hyperbola, the relation experimentally derived by Svedberg.

Utilising his values for amplitude, viscosity and time, Svedberg unhappily discovered that the revised Einstein formula resulted in figures somewhat smaller than his own experimental ones, though the ratio of the predicted and experimental values of A remained more or less constant. The comparison with Smoluchowski's formula leant itself to closer agreement and, according to Svedberg in his 1907 publication, there subsequently followed a correspondence between Svedberg and Smoluchowski over the form of the general theoretical law and the limits of its validity.[84]

Although Svedberg's investigations were to come under fire, both in the conception of a sinusoidal curve and in his alleged inaccuracy in calculating particle size, another of his contributions to the theory of Brownian movement was nevertheless of great importance. As mentioned previously, there was still in 1907 a debate over the question of electrostatic charge in the stability of colloidal solutions, particularly in its causal relation to Brownian movement. While Duclaux disputed this point with Perrin, H. Freundlich insisted—against the arguments of Svedberg—that electric charge alone accounted for the condition of Brownian movement. Svedberg, on the contrary, maintained that the apparent inhibition of Brownian movement upon the addition of an electrolyte was due to a decrease in the amplitude of individual particle movements as the sizes of these particles increased through coagulation. Supporting his viewpoint, he quoted a paper of Picton and Linder describing the formation of a line of coagulation upon the addition of a salt solution to a colloid: 'First appeared a dim cloud of vibrating particles too small to be followed with certainty; *these rapidly increased in size with proportionately retarded movement* until at a certain critical moment the granules began to attach themselves to the free surface of the hydrogel and to one another, and all movement ceased.'[85]

As experimental proof of such an interpretation, Svedberg embarked upon a precise determination of the state of movement of Brownian particles at the isoelectric point, that is, at the point of neutrality in an electrolytic solution. Using gold and silver hydrosols in an electrolytic cell, he added aluminium sulphate to the emulsion, observing both the electrolytic migration (*Wanderung*) and the Brownian movement (*Eigenbewegungen*) of the colloid particles under an ultramicroscope. With the passing of the isoelectric point—in the transition from positive to negative migration—he made especially careful observations with very minute additions of the salt, proceeding so exactly, in fact, that the migration of the particles was practically halted to a standstill. But the Brownian movement continued unabated, as animated as ever. Such an argument was very persuasive of the independence of the movement from electrical influence.[86]

Another series of experiments had been undertaken about 1904 by Seddig at Marburg, and, as with Svedberg, the results were published almost simultaneously with knowledge of Einstein's 1905 article. Seddig was primarily interested in the temperature dependence of Brownian movement, because it was in this that he discerned a convincing proof for the kinetic theory.[87] To this end, he studied a suspension of cinnabar with a vertical ultramicroscope equipped with a microphotographic device. Using an automatic shutter, flashes of light were projected upon the same photographic plate at an interval of 0·1 second, recording two slightly separate images for each moving particle. The photographs were repeated for the same suspension at different temperatures. To his disappointment, the values observed were larger than those to be expected from Einstein's theory. Seddig attributed this to the particles' absorption of heat from the platinum heating coil, resulting in a transmittal of this heat to a layer of liquid encasing the individual particles; consequently the temperature registered by a thermo-element was in fact lower than that which the particles themselves were experiencing. With such an explanation of the discrepancies, the experimental deviations did not undermine his confidence in the theoretical derivations of Einstein nor in the molecular-kinetic theory upon which these were based.

A demoralisation did ensue from the cinematographic studies of Victor Henri at the Collège de France. Employing emulsions of a rubber latex, with granules of one-micron diameter and a density very near that of water, Henri photographed particles with a time exposure of $\frac{1}{320}$ second and at intervals of $\frac{1}{20}$ second. He discovered entirely different and independent trajectories for the granules, very brusque variations in direction, and values suggesting a mean displacement of 0·62 micron in $\frac{1}{20}$ of a second. This displacement was four times larger than the value calculated by substituting Henri's conditions into Einstein's equation, and unlike Seddig, Henri attributed the error not to his own techniques, but to the equation itself.[88] In the next months that followed, most opinion acquiesced to Henri's conviction that, at best, Einstein's derivations were invalid because the application of Stokes' law—a law originally conceived under the conditions of continuity in the substructure of matter—was inapplicable for particles of the order of one micron in magnitude.

The Success of Perrin—'Brownian Movement and Molecular Reality'

In an article in the popular journal *Revue du Mois* in 1908, Aimé Cotton analysed the experimental investigations of Svedberg, Henri, and Perrin in their relationship to the theoretical derivations of Einstein, Smoluchowski, and Langevin. Cotton suggested that the ineptitude of experimental endeavour in verifying the theoretical predictions might belie the credibility of any purely kinetic explanation of the Brownian phenomenon. Perhaps Duclaux's electrical theory did indeed play a necessary complementary role to the cause of the particle motion.[89] The displacements observed by Svedberg were simply too large and his evaluation of the particle dimensions unfortunately inaccurate. In regard to the cinematographic studies of Henri, Cotton conjectured that the large quantity of light necessary for photography might have influenced the mean temperature of the colloid. Perrin's 1908 experiments, Cotton realised, were not a direct confirmation of Einstein's equation, although they implicitly recommended the applicability of Einstein's kinetic and statistical methods to the problem. Wrote Cotton, 'It would be ideal if Monsieurs Henri and Perrin agreed to carry out their photography and measurements *on the very same emulsion*.'[90] Cotton also suggested that Perrin avoid criticism of his 1908 results by using larger microscopic particles with better-known dimensions, a hint which Perrin subsequently followed, as will be seen.

Perrin's own opinion of Svedberg's work was not particularly complimentary when he became familiar with its results in 1909. Perrin dismissed the investigations as adding nothing to the information already embodied in the conclusions of Einstein and Smoluchowski. Svedberg's observed lengths of displacement were six or seven times too large,[91] and, more seriously, Perrin found Svedberg's conception of an oscillatory Brownian movement totally meaningless.

> And what is much more serious, Svedberg thought that Brownian movement was *oscillatory* for ultramicroscopic granules. It is the *wave-length*(?) of the movement which he compares to the displacement of Einstein. It is obviously impossible to verify a theory founded on a phenomenon which, if assumed to be exact, *will be in absolute contradiction with the theory itself*. I might add that, on any scale, Brownian movement does not present an oscillatory nature.[92]

Einstein, too, disapproved of Svedberg's treatment of the subject, writing to Perrin that the mistakes in Svedberg's method of observation, as well as in the manner of theoretical treatment, had become immediately evident to him:

> The mistakes in Svedberg's methods of observation and also in the theoretical methods of treatment had become clear to me at once. I wrote at that time a small correction which seemed to me to correct the worst errors, because I could not resolve to impair for Herr S[vedberg] the great pleasure in his work.[93]

The lack of agreement between Seddig's observations and Einstein's theory was not available in the literature regularly read by Perrin, although he shortly became acquainted with it. Criticisms of Seddig's results were directed at the implicit emphasis upon the influence of viscosity rather than of temperature. As Perrin later remarked, the role of viscosity in inhibiting Brownian activity was seven times greater than that of temperature, and investigations of a strictly temperature-displacement relationship were not particularly useful. One could never really even affirm the influence of temperature from mere inspection, if the viscosity did not diminish with an increase of temperature.[94]

The experiments which did dominate Perrin's musings on the

validity of Einstein's (and Smoluchowski's) derivations were those of Henri and—in retrospect to the situation of 1908, when Henri's first attempt to determine N from Einstein's equation proved unfavourable—Perrin later reminisced,

> I recall the point because I had been sharply struck by the ease with which those physicists the most attached to the kinetic theory were prompt to concede that Einstein's theory might contain some unjustified hypotheses. This proved to me how limited, at root, is the credit we give to theories, and that even those who construct them see here instruments of discovery rather than true demonstrations. Although won over by the general doubt, I tried to measure the agitation of precisely measured granules which I had prepared, and immediately it was apparent that the theory of Einstein was verified.[95]

So it occurred that Perrin initiated in 1909 attempts to verify Einstein's displacement equation, fully expecting negative results. In contrast to his predisposition, he viewed an approximate agreement. He happily observed that 'This impression was more and more confirmed as a larger number of points further eliminated the irregularities of statistics.'[96]

As has been mentioned, the initial researches undertaken in Perrin's laboratory to test Einstein's 1905 equation fell to Chaudesaigues, a student completing the requirements for his doctorate. Henri's negative tests of the displacement equation, as well as his studies linking Brownian movement to electric charge, formed the focal framework for Chaudesaigues' paper. First attacking Henri's insistence that the addition of a coagulating agent inhibited Brownian movement before coagulation actually commenced,[97] Chaudesaigues established that the addition of traces of acid altered Brownian movement only very close to the walls enclosing the emulsion. Then, citing the behaviour in water and aqueous sugar solutions of *caoutchouc* granules, Chaudesaigues announced that his observations conformed to Einstein's formula in regard to the influences of particle radius, of time and of viscosity suggesting a mean value of Avogadro's number, $N = 64 \times 10^{22}$. 'In short, the formula of Einstein should be regarded as exact', Chaudesaigues concluded.[98]

Following upon Chaudesaigues' success, Perrin compared the relationship of Brownian movements and molecular movements in kinetic theory to the relationship of infrared and ultraviolet wavelengths, both aspects of an integral spectrum of light.[99] Enlarging upon his own earlier experiments on the vertical distribution of emulsions and Chaudesaigues' new successful investigation of the displacement equation, Perrin inaugurated several series of experiments, utilising varying conditions and parameters. With the aid of his pupil Dabrowski,[100] he first conducted a new series with granules of mastic, which formed a white emulsion of particles one-third the density of gamboge and, in this case, had a radius 0·52 micron, large enough for an instantaneous photograph of an entire layer at a time, in confirmation of the exponential height distribution.

In affirmation of Einstein's displacement equation, Perrin followed a method to be repeated through three more series of investigations, with the preparation immersed in water to insure an even and observable temperature and the illuminating beam of light filtered through a glass cuvette of water. Two observers worked together, one dotting the position of a granule in the *camera lucida* (*faisait à la chambre claire les pointes*) at the command of the other, at intervals of 30 seconds or 2 minutes, making four readings on each granule. With the observation of several hundred displacements in the fifth series of experiments with Dabrowski, a value of $N = 73 \times 10^{22}$ was obtained, as compared with the figure $N = 70 \times 10^{22}$ from the exponential height distribution for the same emulsion.[101]

Long before perusing Einstein's 1906 paper treating Brownian movement of rotation, Perrin had observed the irregular gyration of the Brownian particles superimposed upon their translation. He had realised that the theory of equipartition of energy (Maxwell, Boltzmann, Gibbs, *et al.*) required an identical mean energy for rotation and translation, independent of the size of the granule and equal to the mean energy of translation of an isolated molecule. Einstein's derivation provided an exact equation to be tested, and once Perrin had found the theory of translation reliable, he did not delay in firing the mathematics of Brownian rotation at the crucible of experiment.

As the sun's rotation about its axis was first observed through following sunspots, so landmarks were necessary for plotting the circumvolutions of the spherical colloidal granules. To this end, Perrin prepared large, amorphous grains of resin. Instead of quickly precipitating the resin in an alcoholic solution with a large excess of water—as with gamboge and mastic—he allowed the water to slowly rise under the alcoholic solution through a funnel tapered to a point (*par un entonnoir à pointe effilée*). By such a procedure, the granules forming in the zone of mixture of the two liquids could attain diameters of 12 microns before falling to the bottom of the water, where they were separated by decantation. Such granules were transparent balls, perfectly spherical, but with small faults or inclusions providing convenient reference marks for observing their rotation. In order to have the same densities of intergranular water and granules, Perrin experimented with several additives, finally settling upon urea, which beautifully prevented an adhesion of the large particles.

Measuring the distance from the fault to the centre of a granule and observing the rotation at equal intervals of time for spherules of diameter 13 microns, Perrin found values of $N = 65 \times 10^{22}$ and $\sqrt{\alpha^2} = 14{\cdot}5°$ per minute, which compared most favourably with predictions from Einstein's equation of $N = 70{\cdot}5 \times 10^{22}$ and $\sqrt{\alpha^2} = 14°$ per minute.[102]

By 1912, Perrin had carried out ten series of experiments confirming the vertical height distribution and seven series establishing the validity of Einstein's theoretical derivations. One series of investigations was pursued with Niels Bjerrum, applying the generalisations to more viscous fluids with hopes of approaching the rigidity of solids. An emulsion of gamboge in 88% glycerine, with a viscosity 115 times greater than pure water, was studied; the Brownian displacements and distributions (a vertical distribution in which the granules accumulated in this case in the upper layers) were observed and values of N calculated.[103] The last two series were performed by M. Bruhat, under Perrin's direction, with gamboge particles in very cold (−9° C) and very hot (+58° C) water, and the results harmonised quite well with the other series.[104]

Throughout the different groups of measurements, Perrin varied conditions for the viscosity, density, temperature and nature of the granule and of the intergranular liquid. Generally, 3000–4000 granules were counted for the calculations of the exponential distribution, whereas 17 000 were observed in the most precise series (series VII). Several hundred pointings were made for most studies of translation, with 1500 in the most precise. The values of particle-volume ranged from 1 to 48, of viscosity from 1 to 250, of density from +10 to −1 and of Kelvin temperature from 1 to 1·25. For the vertical distribution, he attained an average $N = 70 \times 10^{22}$; for displacement of translation, $68{\cdot}5 \times 10^{22}$; and for displacement by rotation, 65×10^{22}.[105] Of these values he placed the most confidence in the height distribution.[106] In explaining the discrepancies between mean values of N derived from the different methods, Perrin insisted that these were all rather precise measurements if one considers the tremendous interval *a priori* possible for N, either from assuming that the drop in concentration will not be negligible for the minute height of several microns (i.e. yielding $N = 0$), or that all the grains are not assembled in the immediate neighbourhood of the bottom of the preparation (leading to the value $N = \infty$).[107]

Several years before Perrin completed the last of his series of experiments, he utilised his results in a thoroughgoing support of the molecular-kinetic theory, and as further investigations only established the validity of his arguments more convincingly, he successfully erected a finely etched barrier against the forays of the energeticists and other foes of the hypothesis of the discontinuous structure of matter. Two articles, both of which were immediately translated from the French into German and English, correlated Perrin's experimental investigations of 1908 and 1909 with the general gas laws and kinetic theory and cited additional research outside of France which led to the same numerical results from widely different phenomena. The first of these articles, completed before Perrin's confirmation of Einstein's equations, appeared in the *Revue du Mois*, the journal where he had published in 1906 on the discontinuity of matter. The second paper made its début in the prestigious *Annales de Chimie et de Physique* and soon became a booklet in English translation.[108] Besides the experimental proofs of Einstein's theory, the latter also contained a historical account of studies of Brownian movement.

In the *Revue du Mois*, Perrin commenced by discussing the two faculties he deemed most prominent in the good physicist. One, which he termed the 'intelligence of analogies', had resulted in the principles of energetics (thermodynamics), while the second—the power to imagine invisible mechanisms which make visible phenomena immediately calculable—he attributed to atomists. Despite the accomplishments of the second method, Perrin grieved, objections had arisen from scientists accusing the logical construction of such a deductive approach of excessive dependence upon the *a priori*.[109] Yet, in one phenomenon, newly observed and submitted to calculation by Perrin himself, the objections were completely voided. As he phrased it in the *Annales*, 'In short, the examination of Brownian movement alone could suffice to suggest that every fluid is made up of elastic molecules, animated with an eternal movement.'[110]

Further, Perrin wrote, as early as the observations of Léon Gouy, the derogatory effect of a study of Brownian movement upon classical thermodynamics had been suspected, for M. Gouy realised that the phenomenon was irreconcilable with any rigid enunciation of Carnot's principle, while it was harmonious with the statistical approximations insisted upon by Clausius, Maxwell and Boltzmann. To many of the critics of a statistical interpretation, '. . . insofar as they are restricted in letting invisible molecules intervene, it would be possible by denying their existence to believe in the perfect rigour of the principle of Carnot'; now, Perrin cautioned, 'This would no longer be reasonable . . . as this rigour is discovered to be in opposition with a *sensible reality*.'[111]

Before Perrin's investigations of Brownian movement, the most easily accessible of the parameters of the kinetic theory had been the mean molecular velocity, which from the calculations of Joule, Clausius and Maxwell could be figured from the relation $pv = 2/3[Nw]$ (w the mean kinetic energy/molecule). Maxwell also determined the mean free path of a molecule from the viscosity of a gas, and Clausius related this path to molecular diameter and the density of molecules in a unit volume. Loschmidt deduced an upper limit for the volume of molecules from considerations of packing, deriving a minimum limit for N and a maximum limit for the molecular diameter d, while on the other hand, Clausius and Mossotti calculated an upper limit for N and a lower limit for d from electrostatic considerations based on the idealisation of perfectly conducting spheres. Van der Waals's equations, meanwhile, suggested a value of $N = 62 \times 10^{22}$, compared to the previous limits of 13 and 200, while in a novel insight into physical theory, van't Hoff extended the gas laws to dilute solutions, assuming that molecules in solution have the same mean energy as in the gaseous state, because the osmotic pressure exerted on a semi-permeable membrane is the same.

With such an analogy already experimentally demonstrated, Perrin demanded,

> Now, is it probable that a particle even larger and formed itself of several molecules, but still independent in the liquid as if it were a sugar molecule, reacts according to a new law to collisions from molecules which surround it? Will it not behave as a very large molecule? . . . this is regarded as probable.[112]

In fact, says Perrin: in reviewing the animated motion of the Brownian phenomenon, one is irrevocably led to Gouy's hypothesis that each particle is agitated by molecular collisions and that the energy maintained by the particle in these collisions is in its mean equal to that of the molecules themselves. Such an equality is a special case of the theorem of equipartition of energy developed by Maxwell, Boltzmann, Gibbs and Langevin; this special application was itself perceived by Einstein in 1905.[113]

In the 1908 article, Perrin then continues with what he terms a more straightforward and concrete verification than possible through the theoretical use of the hypotheses of statistical probabilities. The purpose of his approach is the evaluation of Avogadro's number from the phenomenon of Brownian movement, and if his value be of the same order as from previous methods, both Gouy and all of molecular-kinetic theory are confirmed. Perrin presents the conclusions of his experiments on the vertical distribution of a colloid and, in the *Annales*, gives an account of the investigations affirming Einstein's formulas of displacement and rotation. Reproducing three drawings obtained by tracing the segments joining the consecutive positions of three granules of mastic at intervals of 30 seconds, he describes how half of the mean square of such segments verifies the translational formula of Einstein. To indicate the merely skeleton accuracy of these drawings in portraying the real path of any one granule, Perrin remarks that, 'If indeed the pointings were made from second to second [instead of at intervals of 30 seconds], each of these rectilinear segments would be found replaced by a polygonal figure of 30 sides, relatively as complicated as the figure [Fig. 4, page 134].'[114]

Citing the values of N obtained from his three methods of observationally studying Brownian movement, Perrin also demonstrates how knowledge of this one constant empowers the observer to know such parameters as the density of molecules per unit volume, the constant of molecular energy ($3R/2N$), the mean kinetic energy of a molecule at a given temperature, the electric charge e, as well as absolute molecular weights and dimensions—all of which can be favourably compared to values calculated by Clausius, Maxwell and others from strict theoretical considerations of the molecular-kinetic theory. In addition, the verification by experiment of Einstein's two equations and the independent observation of a colloidal height distribution, which follows from his theory, beautifully establish the validity of Einstein's methods—

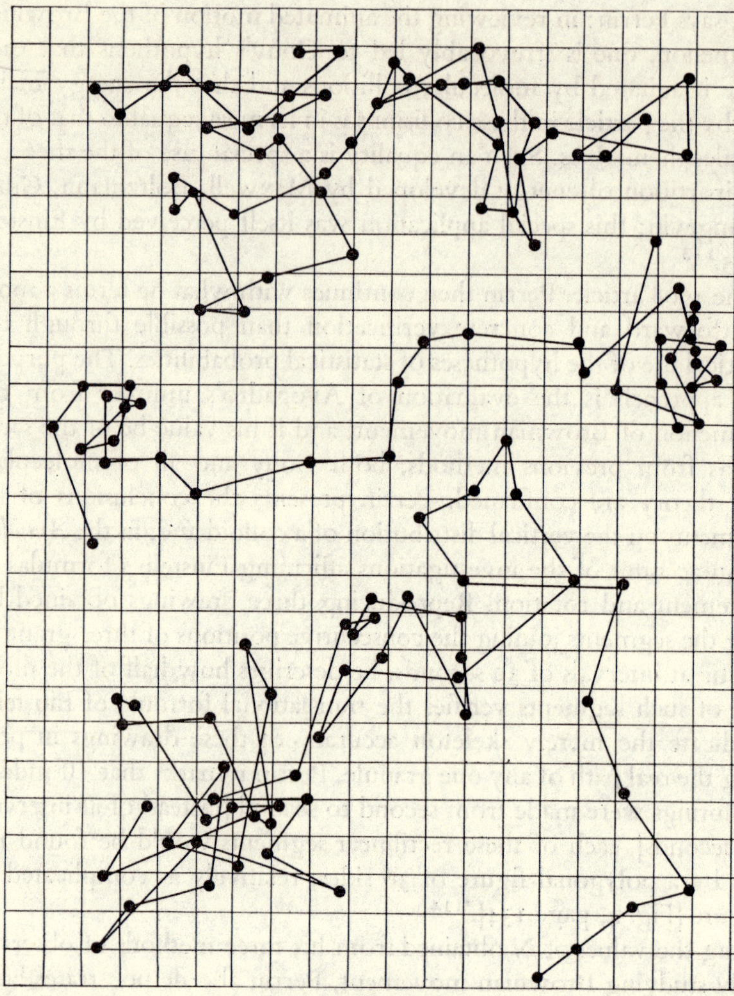

FIG 4[15]

methods which adhere to Boltzmann's statistical approach to thermodynamics and which closely knit together seemingly unrelated properties of liquids. Also clearly accentuated through Perrin's investigations is the validity of the long-disputed hypothesis of the equipartition of energy, upon which Einstein's equations equally depend.

Since Perrin had initiated his studies in order to compare his own experimentally derived values of Avogadro's number (N) to those proceeding directly from kinetic theory, he included in the *Annales* (1909) a table of Ns evaluated from eight different orientations of investigation. These comprised, in addition to the work of Loschmidt, Mossotti and Van der Waals, Einstein's incorrect computations from the diffusion of dissolved substances, Pellat's investigations of the mobility of ions in water, Rayleigh's studies of the blue of the sky, the Cavendish and Millikan's measurements of the electric charge e, de Broglie's and Ehrenhaft's researches on large ions in the air, Rutherford's examination of the emission of α-particles and Planck's and Lorentz's theories of the energy of black-body radiation. De Broglie's and Ehrenhaft's independent results, for example, implied a value of N approximating 64×10^{22}, while different types of studies of α-particles arrived at Ns of 62, 70·5 and 71×10^{22}. These values compared most favourably with Perrin's own 1909 determinations of average Ns of 70·5, 71·5 and 65.[116]

Perrin was most enthusiastic about the novel theories of Planck and Lorentz, and in surveying the discrepancies between his values and theirs (Lorentz: 77×10^{22} and Planck: 61×10^{22}), affirmed that 'One senses that these divergencies will shortly disappear and that the atomistic theories of conductibility and of radiation will lead to the same consequences and be reinforced by one another.'[117] Upon receipt of a reprint of Perrin's 1909 article, Einstein wrote to the author, praising him for a precision in measurement which Einstein had himself feared impossible. Einstein remarked upon the similarity of Planck's and Perrin's values of N, a resemblance which Einstein considered of the greatest import for physics.

> I had thought it impossible to investigate Brownian movement so precisely; it is fortunate for this matter that you have taken it upon yourself. I had not thought a measurement of the pressure to be practicable. In my eyes it was only a nice little game.
>
> A precise determination of the size of the molecule appears to me of the highest importance moreover, because the radiation formula of Planck can be more sharply proved through this than through measurements of radiations. The Planck theory of radiation also leads to a determination of the absolute size of atoms, with *exact* validity claimed. Since the difference
>
> 61×10^{22} (Planck)
> $70{\cdot}6 \times 10^{22}$ (Perrin)

> between the Planck value and your value appears critical to me, it must be said at this time that the Planck formula is apparently only theoretically grounded. Why the question of whether Planck's formula is correct appears so important to me, you can see from the papers which I am sending you with this . . .[118]

In the next few years, Perrin closely followed all refinements of the investigations listed in his compendium of 1909, and in 1911 he reported in an article in *Comptes rendus*[119] on the latest measurements by Millikan, the Cavendish group and himself. He persevered in his own line of research until 1913, when he divined a new possibility for investigating the fine structure of matter by the study of thin films and incorporated these results and further investigations, both within and independent from his laboratory, into an all-encompassing defence of the molecular-kinetic hypothesis.

NOTES

1. This interpretation, confirmed by a careful scrutiny of Perrin's published papers, as well as by the opinion of his son, is a correction of the viewpoint which might be assumed from a mere chronological cataloguing of events. For example, Stephen G. Brush writes that 'Jean Perrin quickly attempted to establish Einstein's theory by experimental test of the displacement formula and of another formula for the vertical distribution of particles in a fluid' ('A History of Random Process. I. Brownian Movement from Brown to Perrin', *Archive for the History of the Exact Sciences*, **5**, #1 (1968), p. 2).
2. Richard Zsigmondy, 'Kolloidchemie', trans. Ellwood B. Spear in *The Chemistry of Colloids* (New York: John Wiley & Sons, 1917), p. 44.
3. Ibid., pp. 47–48.
4. E. F. Burton, 'On the properties of electrically prepared colloidal solutions', *Philosophical Magazine*, **11** (1906), pp. 444–445.
5. Zsigmondy, p. 15.
6. H. Siedentopf, 'On the rendering visible of ultramicroscopic particles', *Journal of the Royal Microscopic Society* (17 June 1903), p. 573.
7. Ibid., pp. 574–576.
8. Aimé Cotton and Henri Mouton, *Les Ultramicroscopes et les Objets Ultramicroscopiques* (Paris: Librairie de l'Académie de Médecine, 1906), pp. 42–43.
9. See Felix M. Exner, 'Notiz zu Brown's Molecularbewegung', *Annalen der Physik* (4), **2** (1900), pp. 843–847.
10. Siedentopf and Zsigmondy, 'Über Sichtbarmachung und Grössenbestimmung ultramikroskopischer Teilchen . . .', *Annalen der Physik* (1903), p. 10.
11. Cotton and Mouton, *Les Ultramicroscopes* . . ., p. 99.
12. Exner, pp. 846–847.
13. Cotton and Mouton, *Les Ultramicroscopes* . . ., pp. 102–103.

14. Duclaux was at this time doing research at the École pratique des hautes études in Paris.
15. Jacques Duclaux, 'Recherches sur les substances colloïdales', *Journal de chimie physique*, **5** (1907), p. 47.
16. Jacques Duclaux, 'Pression osmotique et mouvement brownien', *Comptes rendus*, **147** (1908), p. 132.
17. Jacques Duclaux, *thèse de doctorat*, as discussed in Chapter II.
18. Jean Perrin, 'L'agitation moléculaire et le mouvement Brownien', *Comptes rendus*, **146** (1908), pp. 968–969.
19. Perrin used a number of slightly different equations for the height distribution in the next years, depending upon his purposes. One of his better derivations was published in 'Les preuves de la réalité moléculaire', a report to the Solvay Conference of 1911 and published in *La théorie du rayonnement et les quanta* (Paris: Gauthier-Villars, 1912), pp. 9–10. Here he considered a cylindrical slice of height dh and imagined pistons at either end of the cylinder. On the two faces of the pistons are exercised the pressures p and $(p + dp)$, where $dp = g \, . \, dm$, due to the weight of the slice dm.
    This mass dm is related to the gram-molecule M as dh is to v, the volume of the gram-molecule under the pressure p.
    Hence $dm/M = dh/v$
    and $dp = g(M/v)dh$
    Since $pv = RT$
    then $dp/p = (\mathrm{M}.g.dh)/ RT$
    This gives by integration:
    $$\ln(p_0/p) = (M.g.h)/RT$$
    i.e. each time one ascends the same height, the density is divided by the same number. For an emulsion, one may assume that the pistons are semi-permeable. The force due to weight is the effective weight of the emulsion grains, i.e. their weight minus the force due to the liquid. A gram-molecule of the grains will be $N.m$. The quantity p_0/p may be replaced by n_0/n. Then, one has:
    $\log(n_0/n) = (N/RT)m\ (1 - \delta/\Delta)gh$, where Δ is the density of the granules, and δ is that of the intergranular liquid.
20. Perrin used the relationship $6\pi\zeta av = (4/3)\pi a^3\ (\Delta - \delta)g$, where v is the velocity of fall and ζ the viscosity of the medium. See 'La loi de Stokes et le mouvement brownien', *Comptes rendus*, **147** (1908), p. 475.
21. Jean Perrin, 'L'agitation moléculaire et le mouvement Brownien', p. 970.
22. In a letter from Gouy to Perrin, undated except for the designation '18 Nov.' Judging from the total contents of the letter, it was most probably penned immediately after the publication of Perrin's 1909 memoir in the *Journal de chimie physique*, 'Mouvement brownien et la réalité moléculaire'.
23. In Perrin's 'Les preuves de la réalité moléculaire', in *Les idées modernes sur la constitution de la matière* (Paris; Gauthier-Villars, 1913), p. 20, he lists ten series of experiments undertaken between 1908 and 1911, discussing their principal parameters and acknowledging those of his students who helped him or actually carried out the determinations.

24. Jacques Duclaux, 'Pression osmotique et mouvement brownien', p. 133.
25. See the earlier discussion in Chapter II of Perrin's 1906 paper which was delivered to the Société de Philosophie.
26. Jean Perrin, 'Les preuves de la réalité moléculaire' (1912), p. 169.
27. Jean Perrin, 'Les preuves de la réalité moléculaire' (1913), p. 11.
28. 'Les preuves de la réalité moléculaire' (1912), pp. 171–172.
29. Jean Perrin, 'L'origine du mouvement brownien', *Comptes rendus,* **147** (1908), p. 531.
30. Jean Perrin, 'Mouvement Brownien et réalité moléculaire', *Annales de chimie et de physique* (8), **18** (1909); also trans. Frederick Soddy as *Brownian movement and molecular reality* (London: Taylor and Francis, 1910), pp, 29–30.
31. 'Les preuves de la réalité moléculaire' (1913), p. 20.
32. *Brownian movement and molecular reality,* p. 31.
33. 'Les preuves de la réalité moléculaire' (1913), pp. 18–19.
34. Jean Perrin, 'Mouvement brownien et constantes moléculaires', *Comptes rendus,* **149** (1909), p. 477.
35. 'Les preuves de la réalité moléculaire' (1913), pp. 18–19.
36. Jacques Duclaux, 'Pression osmotique et mouvement brownien', p. 132.
37. Jean Perrin, 'La loi de Stokes et le mouvement brownien', pp. 475–476.
38. 'Les preuves de la réalité moléculaire', (1913), p. 13.
39. Jean Perrin, 'L'origine du mouvement brownien', *Comptes rendus,* **147** (1909), p. 531. In 1913, René Constantin conducted measurements under Perrin's direction demonstrating that Van der Waals' equation for non-ideal gases applied equally well to non-dilute emulsions. He also showed that above concentrations of 3%, Van der Waals' equation was no longer valid for emulsions.
40. Jean Perrin, 'Grandeur des molécules et charge de l'électron', *Comptes rendus,* **147** (1908), p. 595.
41. Ibid., p. 594.
42. Ibid., p. 595.
43. *Brownian movement and molecular reality,* p. 24.
44. This was confirmed by M. Francis Perrin.
45. p. 104.
46. This, at least, is the opinion of M. Francis Perrin.
47. The first detailed study of the mathematical derivations appeared in the 1913 doctoral dissertation of G. L. de Haas-Lorentz (*Die Brownsche Bewegung und einige verwandte Erscheinungen*, Braunschweig, 103 pp.). Mrs. Lorentz also discusses later investigations and certain problems solved by the methods of Einstein and Hopf. In his article, 'Le mouvement Brownien', *Mémorial des sciences mathématiques*, Fasc. 126 (1954), Paul Levy credits L. Bachelier (*Annales de l'École normale supérieure,* **17** (1900), pp. 21–86 and **18** (1901), pp. 143–210) with having elaborated some of the theoretical derivations later evolved by Einstein. But Bachelier was not aware of their import and applied his results only to speculation at the Bourse and to games of chance. 'Bachelier is still the first to have discovered certain interesting properties of the function $x(t)$, particularly its relation with the equation of heat and the law on which the maximum of $x(t)$ in a given interval depends. His first works are prior to those of Einstein' (p. 4).

48. See the article of Leopold Infeld, 'On the theory of Brownian Motion', *University of Toronto: Applied Mathematics Series,* #4 (1940).
49. Albert Einstein, *Annalen der Physik*, **17** (1905), pp. 549–560. Also in the collection of Einstein's papers *Investigations on the Theory of the Brownian Movement*, ed. R. Fürth, trans. A. D. Cowper (New York: Dover, 1956), hereinafter referred to as Dover, pp. 1–18.
50. Dover, pp. 1–2. In the original German (*Annalen der Physik*, **17** (1905), p. 549), the final sentence from this passage reads: 'Erwiese sich umgekehrt die Voraussage dieser Bewegung als anzutreffend, so wäre damit ein schwerwiegendes Argument gegen die molekularkinetische Auffassung der Wärme gegeben.'
51. As Professor Erwin Hiebert pointed out to me, Ostwald himself stressed colloid studies as one of the most important new areas to be explored by physical chemists.
52. It is somewhat difficult here to understand precisely what Einstein meant by the 'traditional conception of thermodynamics'. According to Gibbs, each component of the liquid, even if consisting of heterogeneous phases, would exert its own partial molal free energy. Einstein could hardly include Gibbs' treatment, then, as part of what he is criticising—nor could he include Planck, van't Hoff and Arrhenius. In his 'Autobiographical Notes', Einstein writes that he was unacquainted with the earlier investigations of Gibbs (and Boltzmann) and that he developed statistical mechanics and the molecular-kinetic theory of thermodynamics independently from them. He also was completely unfamiliar with colloid studies at this time. See the notes, in *Albert Einstein, Philosopher-Scientist*, ed. Paul Arthur Schilpp, Vol. I (New York: Harper and Row, 1959), p. 47.
53. Dover, pp. 5–9.
54. Ibid., pp. 9–12.
55. See Dover, pp. 14–15, for the complete derivation.
56. Ibid., pp. 15–17.
57. Ibid., p. 17.
58. Ibid., p. 18.
59. Albert Einstein, 'Eine neue Bestimmung der Moleküldimensionen', *Annalen der Physik*, **19** (1906), pp. 289–306; in the Dover edition, pp. 36–62. The inaccuracy of this value of N was later analysed correctly by Bancelin, a student in Perrin's laboratory, who called to Einstein's attention that certain of his calculations led to incorrect statements for the energy and viscosity coefficients. Einstein made recalculations, which led to a new value for N, that of $N = 6{\cdot}56 \times 10^{23}$. See 'Berichtigung zu meiner Arbeit', *Annalen der Physik*, **34** (1911), pp. 591–592; in the Dover edition, the corrections are included in the 1906 article, pp. 36–62.
60. Albert Einstein, 'Zur Theorie der Brownschen Bewegung', *Annalen der Physik*, **19** (1906), p. 378; in the Dover edition (pp. 19–35), p. 31.
61. Dover, p. 33.
62. Ibid., p. 27.
63. Albert Einstein, *Zeitschrift für Elektrochemie*, **14** (1908), pp. 235–239; in the Dover edition, pp. 68–85.

64. Albert Einstein, *Zeitschrift für Elektrochemie*, **13** (1907), pp. 41–42; in the Dover edition, pp. 63–67.
65. Regarding Einstein's priority for a theoretical treatment of Brownian movement, Smoluchowski wrote in a letter to Perrin dated the 24th January, 1909: 'Concerning what you want to know about my participation in the establishment of the formula $\Delta_n^2 = (RT/N)(1/3nKP)$, I indulge in recording that the priority is due Einstein (1905), an *auteur* whose fertility and talent has inspired in me a profound respect. It was my fault that I delayed the publication of my researches on this subject until July 1966—researches with which I had kept myself busy since 1900 (following the work of Monsieur F. Exner).'
66. In this equation $\Delta\sqrt{3}$ would be equivalent to Einstein's λ_x, the displacement in the x-axis direction. Hence, in a form akin to Einstein's equation, we have

$$\delta = \sqrt{64/27} \cdot \sqrt{\left(\frac{RT}{N} \cdot \frac{1}{3\pi nP}\right)}$$

See The Svedberg, 'Studien zur Lehre von den Kolloiden Lösungen', *Nova Acta Regiae Societatis Scientarium Upsaliensis*, Ser. IV, **2**, #1 (1907), p. 133.
67. Maryan Smoluchowski, 'Zur kinetischen Theorie der Brownschen Molekularbewegung und der Suspensionen', *Annalen der Physik*, **21** (1906), p. 772.
68. Ibid., p. 763.
69. As Perrin later demonstrated, if carefully observed values, such as those of Exner, were regarded as displacements rather than as velocities, they did harmonise with Einstein's equation.
70. Smoluchowski, p. 764.
71. Ibid., p. 779.
72. Ibid., p. 756.
73. See Infeld, pp. 22–24, for this critical analysis of Smoluchowski's work.
74. Smoluchowski, p. 756.
75. Paul Langevin, 'Sur la théorie du mouvement brownien', *Comptes rendus*, **146** (1908), pp. 533–539.
76. Infeld, pp. 39–40.
77. Studies were also underway on the problem of Brownian movement in gases, a subject which Smoluchowski had suggested for investigation in his 1906 paper. In this area of research, the best-known papers at the time were those of F. Ehrenhaft, 'Die Brownsche Molekularbewegungen in Gasen', *Sitzungsanzeiger der kaiserlichen Akademie der Wissenschaft in Wien, math.-naturw. Klasse*, **4** (1907), p. 72; and Hans Molisch, 'Über die Brownsche Molekularbewegung in Gasen, sichtbar gemacht durch ein gewöhnliches Mikroskop', *Zeitschrift für wissenschaftliche Mikroskopie*, **24** (1907), 97-103
78. The Svedberg, 'Studien zur Lehre von den kolloiden Lösungen', *Nova Acta Regiae Societatis Scientiarum Upsaliensis*, Ser. IV, **2**, #1 (1907), pp. 156–157.
79. Ibid., p. 148.
80. The Svedberg, 'Über die Eigenbewegung der Teilchen in kolloiden Lösungen', *Zeitschrift für Elektrochemie*, **12** (1906), pp. 853–860 and 909–910.
81. The Svedberg, 'Studien zur Lehre von den kolloiden Lösungen' (1907), p. 138.
82. Ibid.

83. See 'Studien zur Lehre von den kolloiden Lösungen' (1907), pp. 143–144.
84. Ibid., pp. 158–160. Smoluchowski suggested that Svedberg use

$$2A = \sqrt{\tau/2} \cdot \sqrt{\left(\frac{RT}{N} \cdot \frac{1}{3\pi kP}\right)}$$

85. Ibid., p. 147. Svedberg uses the original English.
86. Ibid., pp. 148–154.
87. Max Seddig, 'Über die Messung der Temperaturabhängigkeit der Brownschen Molekularbewegung', *Physikalische Zeitschrift*, **9**, #14 (1908), p. 465.
88. Victor Henri, 'Étude cinématographique des mouvements brownien', *Comptes rendus*, **146** (1908), pp. 1024–1026.
89. Aimé Cotton, 'Recherches récentes sur les mouvements browniens', *Revue du Mois*, **5** (1908), p. 739.
90. Ibid., p. 741.
91. Jean Perrin, *Les Atomes* (1913), p. 120.
92. 'Les preuves de la réalité moléculaire', in *La théorie du rayonnement et les quanta* (Paris: Gauthier-Villars, 1912), pp. 203–204.
93. A postscript in a letter from Einstein to Perrin, written at Zurich, the 11th November 1909.
94. Jean Perrin, *Les Atomes* (1913), p. 121.
95. 'Les preuves de la réalité moléculaires', in *Les idées modernes sur la constitution de la matière* (Paris: Gauthier-Villars, 1913), pp. 30–31. Perrin follows this remark by stating his view of the role of mathematical theory in science: 'I scarcely need to say that there should not be seen in these lines a criticism of the methods of mathematical physics. These methods have often been reproached for insinuating into the course of calculations hypotheses which easily elude even the most lucid minds. This is possible, but I would not be surprised if the great power of these methods as tools of research and invention was precisely in introducing with proper logic hypotheses which would not have been imagined without their aid.'
96. Jean Perrin, 'Mouvement brownien et molécules', *Journal de Physique* (1910), p. 32.
97. Victor Henri, 'Influence du milieu sur les mouvements browniens', *Comptes rendus*, **147** (1908), pp. 62–65.
98. Chaudesaigues, 'Le mouvement brownien et la formule d'Einstein', *Comptes rendus*, **147** (1908), p. 1045.
99. Jean Perrin, 'Mouvement brownien et moléculaire', *Journal de Physique* (1910), p. 33.
100. Ibid., p. 27, Perrin writes that '. . . upon the friendly insistence and with the aid of Monsieur Dabrowski, I repeated the measurements for granules of mastic'.
101. Jean Perrin and Dabrowski, 'Mouvement brownien et constantes moléculaires', *Comptes rendus*, **149** (1909), pp. 477–479. As Perrin explained in 'Mouvement brownien et réalité moléculaire', one could consider the displacements with respect to two rectangular axes, but this was not necessary. '. . . the granules which I prepared were pointed *à la chambre claire*, the microscope being vertical, which gives horizontal projections of the displacements. Working on cross-

ruled paper, one has the projections directly on two rectangular axes of the diverse segments obtained, but it is useless to measure them, for the sum of the squares of these projections is equal to the sum of the squares of the segments, so that in order to have the mean square of the projection on an axis, it is sufficient to measure these segments one by one, to calculate their squares, and to take half of the mean of these squares.' See the *Annales* or the *Oeuvres Scientifiques de Jean Perrin*, pp. 214–215.

102. Jean Perrin, 'Le mouvement brownien de rotation', *Comptes rendus*, **149** (1909), pp. 549–551.
103. Jean Perrin and Niels Bjerrum, 'L'agitation moléculaire dans les fluides visqueux', *Comptes rendus*, **152** (1911), pp. 1569–1571.
104. 'Les preuves de la réalité moléculaire' (1913), p. 20.
105. Throughout his work on Brownian movement, Perrin's values of *N* were consistently too high, though the imprecision was due to the nature of the methods themselves, rather than to any failure on his part.
106. Jean Perrin, 'Les grandeurs moléculaires (nouvelles mesures)', *Comptes rendus*, **152** (1911), p. 1380. Perrin carried out only a few studies of rotation.
107. Jean Perrin, 'Mouvement brownien et molécules', *Journal de Physique* (1910), pp. 27–28.
108. 'Peut-on peser un atome avec précision?', *Revue du Mois*, **6** (1908) pp. 513–538, trans. in the *Zeitschrift für Electrochemie*, (1909), pp. 269; and 'Mouvement brownien et réalité moléculaire', *Annales de Chimie et de Physique*, 8e serie, **18** (1909), pp. 1–114, trans. Frederick Soddy as *Brownian movement and molecular reality* (London: Taylor and Francis, 1910), and trans. in *Kolloidchemische Beihefte*, **1** (1910), p. 221. The latter is also in the *Oeuvres Scientifiques de Jean Perrin*, pp. 171–239.
109. *Revue du Mois* (1908), pp. 513–514.
110. *Oeuvres Scientifiques*, p. 176.
111. Ibid., pp. 174–175.
112. *Revue du Mois*, (1908), p. 527.
113. Ibid.
114. *Oeuvres Scientifiques*, p. 217.
115. Ibid., p. 218.
116. Ibid., p. 237.
117. *Revue du Mois* (1908), p. 533.
118. Unpublished letter from Einstein to Perrin, Zurich, 11 November 1909.
119. Jean Perrin, 'Les déterminations des grandeurs moléculaires', *Comptes rendus*, **152** (1911), pp. 1165–1168.

Chapter Four

Denouement

The publication of Jean Perrin's researches on Brownian movement from 1908 through 1911 bombarded a community of physicists and physical chemists already immersed in a deluge of radically new experimental information and exceedingly puzzling, but undeniably iconoclastic physical theories. Perrin had himself participated in the first elucidation of the mysterious characteristics of the whole phalanx of rays which had moved into the spotlight in the 1890's. These, of course, included cathode rays, X-rays, radioactive emanations, and even the more illusory N-rays and black light. By 1911 the researches of the Curies, J. J. Thomson, and Ernest Rutherford, with their coteries of associates, had produced a splendid, but nonetheless still somewhat inscrutable array of facts and theories.

Similarly, the chimera of radiation theory had appeared well over a decade earlier, and an accompanying labyrinth of careful researches and imaginative mathematics had been hesitatingly laid out by Rubens and Kurlbaum, Wien, Lorentz, and Planck, climaxing in the awesome enigma of quantum theory. Equally puzzling was the question of specific heats and the laboured attempts to explain the rapid diminution of specific heat with temperature decrease. This problem involved the still questionable assumption of Maxwell's theory of equipartition of the energy of molecules. Following Einstein's paper of 1907, it seemed to hinge as well upon Planck's hypothetical quanta and Boltzmann's generally misunderstood probability theory.

Perrin's investigations were, first of all, a solution to a phenomenon which had been observed and pondered for almost a hundred years. As Stephen Brush writes, in a recent article discussing the history of Brownian movement, one can, in fact, hardly find a better example in the history of science of the complete failure of experiment and observation, unguided by theory for an extended number of years, to unearth the rather simple laws governing a phenomenon.[1] Certainly Perrin's detailed experimental researches and his clear elucidation of the problem of Brownian movement seem in themselves a marvellous achievement, and his first lengthy memoir, 'Mouvement Brownien et Réalité Moléculaire', appears an excellent demonstration of the methodology and precision which particularly characterise successful experimental endeavour.

Yet, as is evident from the very title of that memoir, as well as from the previous detailed discussion of the early work and interests of Perrin, the riddle of Brownian movement hid a most important key to a number of the major problems which had been confronting chemistry and physics for the last hundred years and which, by the end of the first decade of the twentieth century, seemed threatening to rip a once magnificent façade, now niched and blemished from the battles of the energeticists and atomists, completely asunder. In the hands of Perrin, the one weapon of the phenomenon of Brownian movement could become all things to all people: a visual illustration of the truth of kinetic theory, a demonstration of the non-absolute probabilistic interpretation of Carnot's principle and of the law of entropy, an implicit example of the equipartition of energy of molecules, a microscopic unit epitomising the adequacy of mechanical models and the fruitfulness of the imaginative, hypothetical, deductive approach in solving scientific questions and in building theories, and, most important of all, the lively, if not living, proof of the discontinuity of matter and the reality of molecules. Mach's contention that the very definition of atoms and molecules rendered them both a contradiction of experience and an experimentally unverifiable entity seemed suddenly transformed into an extremely vulnerable critique, as did Duhem's misgivings about the convenience with which physical phenomena could be derived from any corpuscular hypothesis. Though Berthelot's demand to actually *see* a molecule or atom had not been strictly met, the reality of the corpuscles seemed no longer an act of faith.

These, at least, were the advantages of the new investigations of Brownian movement, as Perrin optimistically viewed them, and, as will be seen, he was not far wrong in his happy confidence about the reception and denouement of his researches relating Brownian movement to the profound question of molecular reality. The immediate reaction was favourable, with only a few criticisms voiced, and the next few years offered a succession of platforms for the propagandisation of the proofs of the kinetic theory and corpuscular hypothesis. On these occasions, Perrin's enumerations of different experiments included his own current researches, which continued to delve into the realm of the atom, and especially his work on Brownian movement. These podiums included several conferences, both national and international, as well as the publication both of articles and of his enormously influential *Les Atomes* (1913). The response which he met laid the foundation for the resolution of the energetics-atomistics controversy of the past and aided in the clarification of the major issues to be confronted in the future.

Response to 'Brownian Movement and Molecular Reality'

After the publication in *Annales de Chimie et de Physique* (1909) of the 114-page memoir 'Movement Brownien et Réalité Moléculaire', Perrin sent reprints to a number of individuals whom he thought would be most interested, while others responded simply upon their own perusal of the journal. Thanking Perrin for his copy of the memoir, H. A. Lorentz wrote that he was already familiar with the results in the *Comptes rendus*, 'qui comptent parmi les plus beaux de la physique moderne',[2] and H. Kamerlingh Onnes wrote of his admiration for the 'mémoire classique'.[3] That the school of workers at Berlin quickly became familiar with Perrin's work was attested by his being elected an honorary member of the faculty at the Physikaliches Institut der königlichen Universität in 1910,[4] while the response of the English school was exemplified by J. H. Poynting's praise for the 'beautiful and most important experiments on the Brownian movements'.[5] Within France, Branly immediately commented upon the importance of the application of the kinetic theory to the behaviour of emulsions,[6] and Émile Picard wrote with something of a little venom towards the energeticists, 'If things exist when we can make them correspond to numbers, there is no more doubt about the existence of molecules.'[7]

The importance of Perrin's methodology was also noted in correspondence from a number of experimenters, especially those interested in colloid chemistry. As early as the fall of 1909 he was informed that his methods for determining the radius of colloid particles and for calculating Avogadro's number had been applied by workers in Utrecht to submicroscopic agglomerations of casein, and that a value of 62×10^{22} had been calculated for *N*.[8] Perrin's own values for Avogadro's number were also utilised by researchers in other fields. Keesom, for example, wrote that he had used Perrin's value for N ($N = 70{\cdot}5 \times 10^{22}$), culled from an article translated in the *Physikalische Zeitschrift*, in his own researches.[9]

Many of those scientists who had themselves agonised over the explanation of Brownian movement also quickly responded to Perrin's publications. William Ramsay, discovering that Perrin was apparently unaware of Ramsay's own endeavours in this area in previous years,[10] wrote to enlighten him as well as to discuss other historical precedents in the problem.[11] Léon Gouy had, of course, been well known to Perrin as an earlier investigator of the problem and was apparently a recipient of reprints from the first articles by Perrin in the *Comptes rendus*,[12] as well as an enthusiastic reader of subsequent articles.[13] Reflecting some of the turmoil then enveloping the minds of physicists, Gouy requested some enlightenment on Perrin's application of the equipartition theory and his utilisation of the equality of energies of translation of gaseous and liquid molecules and of large particles. 'Is it', Gouy asked, 'one of those postulates in our thought that is verified particularly by its consequences? or is it a true theorem of kinetic theory? and without this case, where is the demonstration?'[14] Further, Gouy queried, with awareness of the new concepts of atomic substructure which were already being bandied about, how would the internal movements of an atom affect both the observed and theorised actions of Brownian movement?[15]

Although Theodor Svedberg apparently did not correspond with Perrin, his thorough familiarity with the current bibliography on the topics of interest to him insured his carefully following the notes and memoirs of the French scientist. In publishing a lengthy book in 1912 on the various proofs for *Die Existenz der Moleküle*, Svedberg claimed priority for himself as the first individual to present quantitative confirmation of the molecular-kinetic theory of Brownian movement. Discussing the investigations of Perrin, Chaudesaigues and Dabrowski, he then lauded Perrin for his excellent proof for the correctness of the molecular-kinetic conception of Brownian movement and for his overpowering evidence for the real existence of the molecule. ('The measurements carried out by Perrin and his students give an excellent proof for the correctness of the molecular-kinetic theory of Brownian movement and, along with that, for the real existence of the molecule.')[16]

Langevin, of course, was a close friend of Perrin and had aided him in the development of the application of laws of probability to the phenomenon of Brownian movement.[17] Smoluchowski, on the other hand, was sent a copy of Perrin's 1909 memoir and replied immediately in January of that year with nothing but praise and emphasis especially upon the possibilities for the computation of Avogadro's number in Perrin's work:

> I am eager to congratulate you, particularly on your principal discovery —this beautiful method of evaluation of the number *N*. I have dreamed, my friend, of putting in evidence the law of distribution of particles (e.g. 33 in my work) in an experimental fashion, but I feared the experimental difficulties, particularly currents of convection and inequalities of the size of the particles; and I scarcely suspected that such exactitude could be attained in these measurements as would permit the determination of *N*.

Smoluchowski attributed Perrin's resolution of the experimental difficulties to his method of microscopic observation of the centrifuged solution, 'a method as simple as it is ingenious'.[18]

The third principal in the theorisation of Brownian movement, Einstein, apparently first wrote to Perrin in late 1909, although the two did not meet until the occasion of the first Solvay Conference in 1911.[19] Einstein himself wrote that he had thought it impossible to investigate the Brownian movement with the precision achieved by Perrin, that he had considered the measurement of the osmotic pressure completely impracticable[20] and he was most pleased with the physical chemist's researches.

Einstein was especially interested in radiation theory at this time and recognised the importance of Perrin's work not only as verification of the molecular-kinetic theory and of the probabilistic interpretation of physical problems, but also as a possible aid in the confirmation of Planck's quantum theory. As discussed earlier, Einstein felt that a determination of molecular size was of the greatest importance since Planck's work could be more readily proved through this means than through measurements of radiation as such. Einstein also remarked that Planck's value for *N* was somewhat lower than Perrin's average value, but Einstein assumed Perrin's value was more accurate since it was experimental.[21] When subsequently Perrin's student, Bancelin, wrote to Einstein that his measurements of emulsions of mastic particles indicated that a coefficient in Einstein's equation for the diffusion of a sugar solution was experimentally incorrect,[22] Einstein was immediately concerned about the long-range implications of this discrepancy.

Having just completed two papers on radiation theory with Ludwig Hopf,[23] Einstein turned to Hopf with the news and with the contention of the problem's importance, as the question of the dissolved molecule's volume was involved.[24] Having spent several days fruitlessly poring over his calculations, he prevailed upon Hopf to look at them, pointing out that if the mistake were not in the calculations, it must mean that the volume of the particles which Perrin had prepared for Bancelin was larger under the conditions of suspension than Perrin believed, and this might weaken the validity of other of Perrin's experiments.[25] Hopf did indeed discover an error in differentiation on Einstein's part, and there was a formal correction in the *Annalen der Physik* in 1911.[26] Writing Perrin again shortly after Hopf's success in discovering the error, Einstein suggested that Perrin nevertheless check his own measurements of the mastic particles, as there still remained a slight discrepancy.[27] From Perrin's following papers, it appears that this error was minute and did not affect any of his own results.

Once his experimental researches had been recognised to be of distinct value by some of the leading physicists of the countries of Western Europe, these experimental results of Perrin from the years 1908 through 1911 were almost immediately translated into other languages for a more general consumption. Frederick Soddy, for example, wrote of the 1909 memoir to Perrin that he was 'anxious that it should become available to other English workers in science: it touches so many separate branches and knits them together in one harmonious whole'.[28] Accordingly the small book, *Brownian Movement and Molecular Reality*, translated by Soddy from Perrin's long paper of 1909, appeared in London in 1910; a German translation of the memoir appeared in the *Kolloidchemische Beihefte*[29] in the same year.

The new interest in Brownian movement and the successful experiments of Perrin had already been touched upon in English journals, such as the *English Mechanic and World of Science*, which were aimed at the layman of science and at scientists concentrating on other specialities. Around the early part of 1909, for example, Fournier d'Albe wrote that, with the renewed experimental investigations of Brownian movement, 'We are one stage farther towards the visible molecule, which becomes less of a speculation and more of an actual visible and tangible fact. . . . Already the large-scale system of thermodynamics to which we are accustomed shows signs of splitting up and a Maxwellian demon peeps through every crevice.'[30] Discussing the experimental and theoretical studies of Brownian movement in two different articles, d'Albe summarised the state of affairs with the affirmation that Einstein, Ramsay, Langevin, and Perrin were among the most progressive physicists of the day who had verified the existence of Maxwell's demon.[31]

A similar discussion, but one with much more command and finesse, appeared in the influential German journal, the *Jahrbuch der Radioaktivität und Elektronik* in 1909.[32] Here, Stephen Jahn, a physical chemist at the Technische Hochschule of Charlottenburg, grappled with the question of the legitimacy of the atomic theory, especially in the light of the fact that Ostwald had recently attempted to demonstrate that the stoichemetrical laws of chemistry could be derived from chemical fact without resort to an atomic hypothesis. These stoichemetrical laws had, of course, traditionally supplied the atomic theory with one of its clearest meanings and firmest supports. If Ostwald's critique were accepted, there remained then as bulwarks for the atomic theory only the stereochemical laws of chemistry and the great complex of facts of the kinetic theory, neither of which had heretofore provided a direct, visible proof of the discontinuity of matter. But Jahn continued, in the last few years, immense progress had occurred through a new understanding of the old phenomenon of Brownian movements, and after discussing the theoretical and experimental work that climaxed in the experimental studies of Perrin and his co-workers, Jahn wrote that it appeared that proofs now existed for the molecular-kinetic conception. Through the interlinking factor of Avogadro's number, the physical atom would become one with the chemical atom.

> It is asserted accordingly that the nature of Brownian molecular-movement is cleared up and is seen as an immediate consequence of the operation of collisions of the moving molecules. Even if it has not also succeeded in making evident the movement of individual molecules themselves, the direct observation of the continuous movement of very large molecular complexes according to the present results of experiment and calculation still means such an approach to the realisation of the molecular-kinetic conceptions, that it approaches a direct full proof of the theories. Chemical atomistics is, however, so closely connected with this theory through the law of Avogadro, that it also appears to be proved.[33]

The German interest in Perrin's work and in its multiple implications was also reflected in the publication of Werner Mecklenburg's book *Die experimentelle Grundlegung der Atomistik* in 1910.[34] Mecklenburg, a chemist who was especially interested in proofs against energeticism, wrote to Perrin in early 1910 requesting more detailed information on Perrin's investigations of Brownian movement, 'la preuve la plus exacte et la plus valeuse de la théorie cinétique'.[35] In his publication of 1910, Mecklenburg then wrote that kinetic theory, colloid chemistry, and the discovery of the electron all confirmed the atomic-molecular theory with its fundamental tenet that apparently homogeneous structures are in fact inhomogeneous. Roughly a third of his book was devoted to an historical discussion of each of these areas of scientific pursuit, with the conclusion that the future of energetics, which had dominated physics for several decades, now lay in the construction of a refined atomistics.[36]

Of far greater interest and of more profound influence were the immediate reactions of the three masters of the so-called 'Leipzig School' of physical chemistry, Walter Nernst of Göttingen, Svante Arrhenius of Stockholm and the venerable Ostwald of Leipzig.[37] Nernst responded to the novel confirmation of the kinetic theory in his sixth edition of the *Theoretische Chemie* (1909) by inserting a new section on kinetic theory, in which he discussed Brownian movement in relationship to the limits of validity of the second law of thermodynamics. He concluded by conceding that: 'Considering such widespread confirmations of the conception which the kinetic theory supplies us concerning the world of the molecule, it must be conceded that this theory has lost its hypothetical character.'[38]

Arrhenius had viewed the tribulations of the kinetic theory at the hands of those who regarded thermodynamics as 'la pure verité' and those who, like Kolbe, opposed the mapping of the topography of molecules by stereochemical formulas.[39] In 1911 he published an analysis of the current state of molecular theory. In the researches on cathode rays and radioactivity and in the theory of black-body radiation he saw convincing evidence for the truth of the ancient hypothesis, but if these were not sufficient, then 'By good fortune, another different method . . . was found' in the theory of Einstein and Smoluchowski and in the investigations of Svedberg and Perrin. 'It was left to M. Perrin to draw extremely important consequences from an exhaustive examination of the properties of these small particles. . . . You are doubtless familiar with this masterly work executed with a gallic elegance . . .'[40]

The third of the magistral trio, Wilhelm Ostwald, who, in his Ingersoll lecture at Harvard in 1906, had still affirmed that 'atoms are only hypothetical things',[41] first wavered in his unmitigated hostility to the atomic-molecular hypothesis in a 1908 review of Theodor Svedberg's *thèse,* the *Studien zur Lehre von den kolloiden Lösungen.* In the measurements of young Svedberg on the Brownian movements, which supported (Einstein's) theoretical determinations founded upon the kinetic theory, Ostwald recognised the long-sought experimental proof for the kinetic theory. His own past rejection of the theory had been based on the lack of any such proof, Ostwald wrote, and his vehement protest against the validity of the discontinuous hypothesis had been fully justified until irrefutable evidence supporting the atom's reality was indeed presented. His had been a role of criticism and rational inquiry necessary to scientific progress.[42]

A more complete recantation of his prior interdiction of the molecular hypothesis soon appeared in Ostwald's preface to the fourth edition of his *Grundiss der physikalischen Chemie* (1908). This edition was marked by the addition of several new chapters dealing with gas ions and radioactivity on the one hand, and with colloids on the other. Pointing out again that his past condemnation of molecular reality had been grounded in the absence of any conclusive and direct experimental proof of the molecular proposition, Ostwald now directed his readers to the work of J. J. Thomson's school and to the conformity between the phenomenon of Brownian movement and the predictions of kinetic theory which had been gleaned from the investigations of a line of researchers, but most conclusively by Jean Perrin.

> I have satisfied myself that we arrived a short time ago at the possession of experimental proof for the discrete or particulate nature of matter—proof which the atomic hypothesis has vainly sought for a hundred years, even a thousand years. The isolation and measurement of gases on the one hand, which the lengthy and excellent works of J. J. Thomson have crowned with complete success, and the agreement of Brownian movement with the demands of the kinetic hypothesis on the other hand, which have been proved through a series of researches and at last most completely by J. Perrin, entitle even the cautious scientist to speak of an experimental proof for the atomistic constitution of space-filled matter.[43]

With this, concluded Ostwald, the atomic hypothesis might now assume its place as a well-founded scientific theory and might enter, too, into the domain of general chemistry.[44] Of Perrin's later publication of *Les Atomes*, he spoke with 'eine grosse Freude'.[45]

Ostwald's son, too, reacted quite favourably to the work of Perrin and to his resolution of the controversy between energetics and atomistics. Wolfgang Ostwald was at this time editor of the *Zeitschrift für Chemie und Industrie der Kolloide*, and it was, in fact, his instigation that resulted in the translation of Perrin's 1909 memoir into the *Kolloidchemische Beihefte*.[46] In that year he wrote to Perrin asking permission to translate the article, expressing his admiration for the investigations. He also assured Perrin that he shared whole-heartedly the desire for a reconciliation of energetics and atomistics[47] and that he himself was currently writing a book which might further that end, by dealing with the relationship of the discontinuous structure of matter to certain surface problems.[48]

The Conferences

Despite this favourable reception from a large number of scientists, Perrin still viewed his task as far from concluded. While other proponents of the molecular viewpoint, such as Eugène Bloch and Perrin's personal friend J. H. Rosny-aîné, published articles for the *Revue du Mois* on 'atomisme dans la physique contemporain', and 'les principes de l'énergétique' in defence of atomism and in praise of Perrin,[49] he himself felt compelled to take both the pen and the rostrum to further bolster the fortunes of kinetic theory and the discontinuous concept. As if in agreement with Abel Rey's formulation that the success of a scientific theory depends in large part upon its ability to be popularised,[50] Perrin responded quite positively to the pleas of journalists like Louis Olivier that 'le grand public scientifique demande la page entière, la page qu'il puisse lire',[51] of the investigations on Brownian movement. Accordingly Perrin composed a lengthy discussion of 'l'agitation moléculaire'[52] which was published in the *Revue du Mois* in 1912, and more immediately, he presented his case at numerous meetings and conferences, both professional and popular, in the intervening years.

The first of these conferences was that of the Société de Physique in April of 1909, just before the publication of the 1909 memoir, and Perrin's disclosure involved essentially the same information and presentation as the memoir, though in less detail.[53] This was followed by a conference at the École professionelle supérieure des postes et télégraphes in 1910,[54] and by another meeting with the Société des Amis de l'Université de Paris in 1911.[55] In these latter two meetings, Perrin enlightened his audiences with a more general history of atomism, admitting, almost in the earlier words of Berthelot, that 'Neither Moschus of Sedon, nor even Democritus of Abdère or his friend Leucippus have left us fragments which suffice for judging what things in their work could have some scientific value.'[56] Only since eighteenth-century studies of fluid pressure by Bernoulli and Euler and nineteenth-century observations of gaseous diffusion, Perrin explained, had the molecular hypothesis been truly viable. Then, before pressing forward with an explanation of recent studies of the phenomenon of Brownian movement and its proofs of molecular reality, Perrin delighted the *amis* with the visual confirmation of the particles' movements by showing the films made by Victor Henri and Comandon.[57]

In addition to the conferences in France, Perrin was hospitably encouraged to participate in a number of other national meetings. Ernest Rutherford, who had quite the investment in the stock of evidence accumulating in confirmation of molecular reality, wrote to Perrin of the forthcoming 1909 meeting of the British Association at Winnipeg: '. . . you would be a great source of strength to our deliberations in Section A', he confided to Perrin, 'and I am naturally anxious to have people who think in "ions" and "electrons". I must congratulate you in your admirable work in determining *N* by a new method.'[58] A second invitation from the English school also resulted in the appearance of Perrin for the vindication of the atomic-molecular theory, as Perrin delivered a lecture at the Royal Institution[59] on Brownian movement in early 1911 upon the request of William Crookes.[60] Later that same year, he was asked by F. Kurlbaum of Berlin to participate in the eighty-third congress of the *Gesellschaft deutscher Naturforscher und Ärtze*.[61]

The two most influential conferences which Perrin was to meet in these years occurred in 1911 and 1912, and the response which he met on these two occasions was by and large indicative of the eventual fate of his researches, as well as of the tenor of the physics of the future. At a time when chaos reigned in the physical sciences, a common meeting ground—both physical and intellectual—was imperative if any normalcy was to be restored to the working conditions of science. These two conferences, the first an international one at Brussels and the second an annual meeting of the physicists of France, sought as goals the clarification at least, the resolution at best, of two decades of confusing and revolution-making data and theory.

The first Solvay Congress of Physics, endowed by the financier Solvay (1833–1922) because of his lively interest in energetics, biology and the social sciences,[62] convened in Brussels in the autumn of 1911 at the Hôtel Métropole. Its international cast, by invitation only, included the most brilliant and controversial figures in physics: Curie, Brillouin, de Broglie, Einstein, Hasenöhrl, Jeans, Knudsen, Langevin, Lindemann, Lorentz, Nernst, Kamerlingh-Onnes, Planck, Poincaré, Rubens, Rutherford, Sommerfeld, Warburg, Wien, Goldschmidt, and Perrin. Rayleigh and Van der Waals were designated as *membres adherents*. Lorentz served as president and his opening discourse pinpointed the issues and purposes which were at stake in the five days of meetings.[63]

The momentous questions confronting the conference, Lorentz asserted, were those of the very principles of mechanics and the most intimate properties of matter. They involved the fundamental equations of electrodynamics and traditional ideas on the nature of the ether, 's'il est encore permis d'employer ce mot'.[64] At this hour, he continued, physicists remained dissatisfied with the kinetic theory of gases which had been extended little by little to fluids, to dilute solutions and to systems of electrons. In this state of affairs, the beautiful hypothesis of elements of energy postulated by M. Planck and applied to numerous phenomena by Einstein, Nernst and others had been a precious point of light; all this would be the subject of the congress's discussions.[65] That many of the problems facing the gathering in 1911 were based upon an impasse in the face of the atomic-molecular hypothesis—a deadlock which was many decades old—was also noted by Nernst. Recalling the international meeting of chemists at Karlsruhe in 1860, Nernst remarked, 'That Congress was likewise called in order to study the one fundamental question of atomistics.'[66]

Twelve papers were presented in these five days, dwelling principally upon the subjects of radiation and quantum theory, specific heats and kinetic theory. Jeans read Rayleigh's paper on radiation theory, and Rubens and Warburg presented their experimental data in verification of Planck's quantum theory, which Planck, in turn, elaborated to the intrigued and sceptical group of physicists. Nernst dealt with the application of quantum theory to solids, while Kamerlingh-Onnes discussed his studies of low temperature. Sommerfeld discoursed on the physical significance of Planck's constant and the theory of X-rays, while Einstein treated specific heats. Jeans, too, dealt with the theory of specific heats, although studying it from the standpoint of Maxwell and Boltzmann, rather than from quantum theory, while Langevin presented his work on the kinetic theory of magnetism, and Knudsen reported on the application of kinetic theory to the experimental properties of gases.[67]

Perrin's paper, *Les preuves de la réalité moléculaire*, was scheduled for discussion a few days after the Congress had been in session, and the full paper was the most detailed of any he ever devoted to the topic on any one occasion. He realised without reservation the critical importance of a convincing presentation if all doubts of the validity of the atomic-molecular hypothesis were to be quelled among his *confrères*. Accordingly, he detailed the history of the problem, adding Ramsay's information about Buffon, Spallanzani, and Jevons, as well as an explanation of the more relevant conceptions of Ramsay, Svedberg, Seddig, Smoluchowski, and others whom he had not previously credited. In addition to the usual seats of action of the Brownian phenomenon, he commented that he himself had even observed it for particles supported by the black spots of soap bubbles, that is, for particles plunged simultaneously in two different fluids. He had also viewed it on thin films of thickness (6 millimicrons) exceedingly fragile in comparison to the diameter of the suspended globules themselves.[68]

After detailing his own investigations and explaining in depth their relationship to the kinetic theory, to the theory of equipartition of energy, to the probabilistic interpretation of Carnot's principle, and so on, Perrin concluded by summarising those other proofs besides his own for the atomic-molecular hypothesis, all of which were resulting in converging values for Avogadro's number and other molecular constants, such as molecular volume and electric charge.

In response to Perrin's presentation, there emerged only one critical query, when Lindemann asked if Perrin had allowed the possibility of a radiation pressure from his source of illumination adversely affecting his results. Perrin replied that a radiation pressure would be negligible, both because the illumination was instantaneous, in order to prevent a modification of the particle distribution, and because the experimental results did not demonstrate any dependence on the intensity of light.[69]

The further discussion which ensued centred upon current measurement of the value of the electrostatic charge e, in particular the erroneous values of Ehrenhaft, derived from studies of the movement of particles in air.[70] There was, then, no real opposition voiced against Perrin's conclusions, so effective was his presentation, and the Solvay Conference proved one of the most important platforms for the elaboration of his experimental data and the dissemination of his viewpoint of kinetic theory and molecular reality. One suspects, too, that the very juxtaposition of his incisive, straightforward, and logically reasoned paper against the more confusing and perplexing reports on radiation and quantum theory created the emotional feeling that at least one major problem had already been definitely resolved.

The following year, Perrin presented substantially the same paper to the annual meeting of the Société française de Physique, with the additional information that his student Bruhat had experimentally verified the expansion of the height of a column of Brownian particles for high temperatures and its compression for lower temperatures.[71] Perrin's was the first paper presented at the conference, whose members included Langevin, Edmond Bauer, Eugène Bloch, M. A. Blanc, L. Dunoyer, Mme Curie, A. Debierne, Pierre Weiss, and Henri Poincaré. At this conference attention was centred on the evidence for a continuous structure of matter, including Planck's quantum theory, with subjects of discussion including the electronic theory of metals and electromagnetic dynamics, ionisation by collision, radioactive radiations and transformations, and the magnetic moment of atoms. The conference was concluded by Henri Poincaré, in his discourse, *Les rapports de la matière et de l'éther*.[72]

Poincaré, of course, had harboured a number of reservations about the kinetic theory throughout his scientific career, although as has already been discussed, he was one of the first mathematical physicists to realise the potentiality of the Brownian movements as a fruitful experimental approach to a deeper understanding of the substructure of matter: this insight was accompanied, however, by the rather conservative hope that nothing revolutionary would really come of it.[73] But the Solvay Congress of 1911 had convinced Poincaré that an epoch had passed, and in the last of his memoirs before his death he described in no uncertain terms the sharp conflict between traditional theories and new facts, averring that electromagnetic phenomena, of which atoms were now recognised as the cause, could not be represented by differential equations. As Paul Langevin later expressed it, 'At the moment of his own death, then, he marked the end of this period of three centuries during which the admirable instrument of the infinitesimal calculus was set up in the hope that it would allow the enunciation of the laws of the world.'[74] With the advent of the atom and the apparent demise of differential equations, not only did traditional, classical mechanics face a necessity for a face-lifting, but also the principal tenets of energetics and thermodynamics were seriously limited and impaired.

At the conclusion of the 1912 conference of the Société française de Physique, Poincaré contemplated aloud the implications of the set of papers which had been presented and discussed that day, assessing the transformation in thinking which was occurring before their eyes, an intellectual conversion in which the experiments and discussion of Jean Perrin were instrumental:

> A first reflection is sure to strike all the listeners; the long-standing mechanistic and atomistic hypotheses have recently taken on enough consistency to cease almost appearing to us as hypotheses; atoms are no longer a useful fiction; things seem to us in favour of saying that we see them since we know how to count them. . . . The brilliant determinations of the number of atoms made by M. Perrin have completed this triumph of atomism. . . . In the procedures deriving from Brownian movement or in those where the law of radiation is invoked . . . in . . . the blue of the sky . . . when radium is dealt with. . . . The atom of the chemist is now a reality.[75]

Yet this atom, as Poincaré pointed out, failed to fulfil the promise of an ultimate knowledge of things, for it, too now revealed a sub-structure of even smaller elements.

Further Research and Publications

With the completion of the conference of 1912, Perrin still remained resolute in his determination to further espouse the cause of molecular reality and to discover even another experimental verification in its favour. Though a majority of European physicists and chemists seemed converted or confirmed by his endeavours thus far, he was aware of a lingering doubt in the minds of a few—even of his own friend Georges Urbain—and of the apparently inexpungable hostility of an even smaller minority, including Ernst Mach. With that indefatigable verve which had characterised his experimental investigation and his scientific and popular explications from the very beginning of his career, Perrin embarked upon an entirely new train of researches—that of thin film phenomena—as well as upon the composition of what was to prove, by far, his most influential book, *Les Atomes*.

In the first edition of *Les Atomes* in 1913, Perrin included a short discussion of observations of thin films in soap solutions (Reinhold and Rücker) and of oil films on water (Rayleigh and Devaux), displaying the interest which he himself had held since at least the publication of 'Les hypothèses moléculaires' in 1901.[76] Newton and many others before him had observed the small round black spots which appear as a soap bubble is blown thinner and thinner. These black surfaces might be observed quite carefully if the thin film were produced on a fine framework inside a case. Reinhold and Rücker, and, independently, Johonnott, had measured the thickness of the most fragile of these *taches noires* at 6 millimicrons.[77] In investigations of oil films, which Rayleigh had demonstrated might become even thinner than the black spots of soap bubbles, Devaux had very accurately determined the thickness at 1·10 millimicrons.[78] From this evidence, Perrin affirmed, one had direct and incontestable verification of the molecular viewpoint and even of molecular dimensions: 'that the maximum diameter possible for molecules of oil is of the order of a millimicron.'[79]

Beginning in 1913, Perrin initiated his own investigations of the phenomena, discussing various 'Observations sur les lames minces' at the meeting of the Société de Chimie physique in 1913 and carrying out researches in the next few years with René Marcellin and P. V. Wells which resulted in detailed papers in 1918 and 1923.[80] Observing microscopically thin horizontal films of soapy water (5% alkali oleate),[81] Perrin observed that there were different bands of successive (vertically distributed) colours in any one film, and that the stratification of the thin film into these coloured sheets was provoked by the absorption of light (achieved by an addition of fluorescent material) or an increase in temperature.[82]

Measuring the thickness of the stratified bands by the intensity of reflected light ($I \sim \sin^2 2\pi h/\lambda$), with $h =$ the thickness and $\lambda =$ the wavelength of monochromatic light),[83] Perrin found from 1913 on, the minimum mean difference of thickness between any two contiguous bands to be between 4·2 and 5·5 millimicrons.[84] Producing thin films composed of more than a hundred layers, with sharp edges and of uniform thicknesses, Perrin consistently demonstrated that the difference in thickness between any two adjacent bands never fell below a certain value and that this elementary difference was included a whole number of times in each band. Determining the diameter of a film of oleic acid to be approximately 1·9 millimicrons, Perrin was then enabled to suggest that each band seen in the soapy solution contains a layer of solution between two bimolecular layers of oleic acid.[85] As he expressed in his Nobel lecture of 1926, the thin film phenomenon is one where the discontinuous structure of matter can be seen directly,[86] and in later editions of *Les Atomes*, he directed the reader to the newest researches on thin films as further verification of the existence of atoms and molecules.[87]

The publication of *Les Atomes* in 1913 loomed as a milestone in Perrin's own career, as well as in the history of the physical sciences. It marked the alliance of a plethora of superficially unrelated phenomena, as well as the incorporation of the common denominator among them into the normal working assumptions of physics and chemistry. First published in 1913, the book was in its fourth edition in 1914 and was translated into not only English and German, but also into Polish, Russian, Serbian, and Japanese.[88]

In his preface to the book, Perrin discussed the general problem of continuity versus discontinuity, with references to the inevitable discovery of non-homogeneity in almost any superficial homogeneity whether in mathematics (definition of a tangent), biology (the living cell), or physical phenomena (charcoal, colloids). In fact, '. . . the notion of continuity results from a totally arbitrary choice of our attention among the data of experience.'[89] He also enunciates briefly the first and second laws of thermodynamics, alluding cursorily to Ostwald and energetics, and as in earlier essays,[90] comparing two totally different types of intellectual approaches characteristic in physical science—one, the power of perceiving analogies and the other, the imaginative explanation of visible complexities in terms of invisible simplicity. In examining a machine, Perrin explains, 'Certainly we observe as best we can these visible pieces, but we seek also to *guess* what gearings, what *hidden* agencies explain the apparent movements',[91] at least according to that intuitive intelligence by which science has arrived at atomism.

In the following eight chapters of the book, Perrin delineates a great deal more than the tenets and values of kinetic theory, the laws and experimental investigations of Brownian movement, and the theory and researches of fluctuations—the subjects upon which he had primarily focused in the 1909 memoir and in subsequent conferences. Rather, only three chapters are devoted to these topics, and the majority of material is selected from researches in which Perrin was not directly involved, but which further support the hypothesis of a material atom at the core of any given physical phenomenon.

Beginning with a description of the place of atomic theory in traditional nineteenth-century chemistry, Perrin discusses the stoichemetrical laws; substitution, chemical valency, and stereochemistry; diffusion, Raoult's law, Arrhenius' ion theory, and osmotic pressure. After dealing with kinetic theory, Brownian motion, and fluctuations, he turns to the revolutionary quantum theory and its experimental basis in the problems of black-body radiation. 'Every cavity completely enclosed in matter of uniform temperature is full of light in statistical equilibrium',[92] he explains, before expounding Stefan's law, the composition of light emitted by black bodies, and Planck's quantum theory, with its associated values of h and N. Demonstrating the numerous applications of the concept of quanta of energy, he further elucidates the new theories of specific heat and the distribution of matter in a molecule.[93]

The final chapters of *Les Atomes* then deal with what he terms 'l'atome de l'électricité' and the 'genèse et destruction d'atomes', including the series of researches on cathode rays and X-rays, gas ionisations and positive rays; the determinations of electrostatic charge; radioactivity and the radioactive series, X-rays and other atomic projectiles. He devotes an especially careful discussion to the work of J. J. Thomson and to the measurements of Millikan. In conclusion, charting a convergence of determinations of Avogadro's number from the array of diverse phenomena, Perrin writes:

> One is seized with admiration before the miracle of such precise agreements coming from phenomena so different . . . this gives a probability to molecular reality really bordering on certitude. . . . The atomic theory has triumphed. Numerous just a short time ago, its adversaries—finally won over—renounce one after another the suspicions which were legitimate and no doubt useful for a long time [cf. Ostwald]. From now on, that conflict between the instincts of prudence and audacity, whose equilibrium is necessary to the slow progress of human science, will be pursued on other grounds.[94]

The data which Perrin had so triumphantly marshalled in behalf of *les atomes* were the following[95]:

		$N/10^{22}$
Viscosité des gaz (équation de Van der Waals)		62
	Répartition de grains	68·3
Mouvement brownien	Déplacements	68·8
	Rotations	65
	Diffusion	69
Répartition irrégulière des molécules	Opalescence critique	75
	Bleu du ciel	60(?)
Spectre du corps noir		64
Charge de spherules		68
	Charges projetées	62·5
Radioactivité	Helium engendré	64
	Radium disparu	71
	Energie rayonnée	60

Indeed, the future was to bring only the closer precision between measurements of fundamental values which Perrin so exuberantly expected.

The Influence of Les Atomes

The response to the publication of *Les Atomes* was an almost unanimous crescendo of applause. As previously mentioned, Wilhelm Ostwald received a personal copy of the book from Perrin and thanked him for the 'sehr interessantes Buch über die Atome, das ich sofort durch geblättert habe und an dem ich eine grosse Freude gehabt habe. . . .'[96] Lucien Poincaré, who was also a recipient of a gift of *Les Atomes*, ventured that the book marked a high point in the history of physics and the history of science.[97] Jacques Loeb later reiterated the general opinion: 'I hardly need to tell you that your proof of the real existence of molecules will live in science as an epoch-making discovery—and not only in science but in our general philosophic view of the universe.'[98]

Similarly John D. Bernal, fellow of the Royal Society and of Birbeck College, recalled after Perrin's death: 'I myself was a young student when Perrin's book on atoms appeared. . . . It demonstrated the existence of atoms and the possibility of counting them.'[99] Where there had been only speculation and metaphysics, reminisced Bernal, Perrin had established a pragmatic theory, a very real theory. Commenting further on the style and character of the book, Bernal maintained that in England, 'we had been struck by these very clear, very simple, very exact expositions, by this novel method of expressing scientific data. . . .' The imaginative experimental methodology appealed to the English tradition, remarked Bernal, and yet simultaneously the explication of *Les Atomes* breathed into the refined mathematics of the Germans, in all its complexity and abstraction, an aura of reality and comprehensibility.[100]

William Ramsay had already voiced his delight with the writing style of Perrin, the '. . . French style; like many of your compatriots, you possess a facility of exposition, a clarity of illustration that we of the North, less sensitive than you, cannot attain'.[101] Similarly, in his review of the book in the *Journal de Chimie physique* in 1913, Philippe A. Guye, who had in the past moved with discretion in openly supporting kinetic theory and the molecular hypothesis,[102] commended the clarity, simplicity of exposition and elegance in form which characterised all of Perrin's writings.[103] In correspondence with Perrin, Guye further wrote, 'The cause of atomism—which I, moreover, never doubted—has been won.'[104]

Pinpointing the tremendous importance of *Les Atomes* and perhaps focusing upon the crux of the matter, a reviewer in *Nature* commented that

> In these days, when such notable and extensive advances are being made in nearly all fields of physical research, it is extremely desirable that the results which mutually bear upon one another should from time to time be collected together and recorded in more or less popular language. If, in addition, a leading expert can be persuaded to undertake the record, the event of its publication is still more to be welcomed. Prof. Perrin is the ideal author for a book on atoms and molecules. . . . It is not always that the brilliant experimentalist is an equally brilliant exponent, but in the present case it is true, and the book makes fascinating reading.[105]

Indeed, while Theodor Svedberg's book *Die Existenz der Moleküle*, which had emerged from the press one year earlier, was often reviewed in conjunction with *Les Atomes*,[106] its breadth of subject matter, clarity of exposition and potentiality for straightforward reading pleasure all fell far short of that of Perrin's masterpiece. Svedberg had divided proofs for the existence of molecules into two major categories, which he designated as 'multimolecular' appearance and 'paucimolecular' appearances. He then catalogued experimental and theoretical studies of colorimetry, spectrophotometry and diffusion; then of Brownian movement and spontaneous fluctuations of concentration in radioactive solutions and gases. Though detailed and informative, the treatise lacked the more universal appeal of *Les Atomes*' contents and approach, as well as the logically structured analysis of both the traditional and the revolutionary fundamentals of the new physical chemistry, climaxing in a comparison of the numerical constants derived from each.

While the reviewers in *Nature* praised *Les Atomes* for its clarity and simplicity, foreseeing its usefulness to the scientifically inclined general public, J. J. Thomson was well aware that Perrin's was a clarity brilliantly superimposed upon a quite formidable sophistication of data and theory. Such an achievement was not only for the educated public, but might sow the greatest benefits as a source for courses and lectures in physical chemistry. 'Very many thanks for the copy of your book Les Atomes . . .,' he wrote in 1913. 'It has come most opportunely for I am just beginning a course of lectures on the Atom at the Royal Institution and hope to avail myself of the help of your book.'[107]

Indeed, in the next years *Les Atomes* emerged as a popular exposition of the new principles of physics and chemistry, and its approach was consciously imitated by other writers. Within the next decade the contents of new texts of physical chemistry necessarily had to be altered, and as with the text of Edward Washburn, prefaced in its French edition by Perrin and used both in the United States and France, the transformation was quite self-conscious. In regard to the development of his text, Washburn wrote that 'Perhaps the most radical break with the tradition which has prevailed until now in most of the books of physical chemistry, according to the example of Ostwald, is the manner in which atomistics and molecular kinetics are treated.'[108] Instead of explaining these theories in a special chapter as interesting, but unproven hypothetical explications, Washburn considered them as facts established beyond all possible reasonable doubt. Analysing the atomic-molecular kinetics in its essential characteristics and dealing with its multitudinous ramifications, both kinetics and thermodynamics served as the framework for the entire pedagogical development within Washburn's textbook.[109] The influence of Perrin upon Washburn's conceptions is evidenced in his general approach, most obviously in his chapter on Brownian movement and molecular magnitudes, as well as in the fact that it was Perrin who was requested to write the preface to the text's French translation. Analogously, the

Cours de la Chimie pour les Classes de Mathématiques-Spéciales, compiled by Lamirand and Brunold and prefaced by Perrin, employed the molecular framework, once *avant-garde*, now *à la mode* in the 1920s.[110]

The assimilation of Perrin's molecular-kinetic viewpoint and the influence of his specific publications was not strictly scientific in scope, however. The reasons for the impact of his endeavours may be seen in the statement of Pierre Delbet, a member of the Faculté de Médecine at the Sorbonne and a contemporary of Perrin, who had been engaged in medical studies while Perrin was labouring at the École normale.

> I had been brought up in positivist ideas. They never satisfied me completely. The pretension of wanting to impose limits to science shocked me. . . . I passed by a terrible crisis, the memory of which is still painful to me. I then devoured a number of works, without finding peace. When there is so much of reality to understand and the beautiful language of mathematics to reason logically upon ideas, rethinking its thought appeared to me as the most sterile work that a mind could do, and I considered it dangerous when I realised that it could lead to throwing suspicion on science. Moreover, I had arrived at doubting the reality of the exterior world.[111]

Progressively, Delbet explains, he arrived at several philosophical notions which provided him with a modicum of peace, among these that 'le transformisme lamarckien' furnishes an understanding of man and that 'The kinetic theory, in changing the scale of scientific problems, reduces the notion of causality to an arithmetic necessity. Certain knowledge acquired about molecules and atoms has an absolute character.'[112]

Revealing the influence of Perrin's writings and the incorporation of this physical chemist's viewpoint, in particular, into the educated French mentality of the early twentieth century, Delbet writes of the predictions of thermodynamics, of the perception that no reversible phenomena exist and that all occurrences must leave a definitive trace, the imprint of time: 'The principle of Carnot is the principle of evolution, as Langevin and Perrin have said.'[113] In consonance with Perrin's viewpoint, Delbet further condemns the purely energeticist approach, insisting that the dualistic approach is far more fecund, that 'It is equally metaphysical to deny energy and to deny matter', and both denials limit the scope of inquiry of science.[114] As explicative of the nature of matter, he cites the experiments of Perrin on Brownian movement and, only briefly mentioning contemporary researches on viscosities, ions and radiations, asserts that 'Science has thus arrived at knowledge of an absolute nature regarding matter.'[115] Through the insights of kinetic and probabilistic theory, he concludes, the principle of causality is no longer something mysterious; it is the consequence of a 'hasard vrai' expressed in the language of causality.[116]

The intellectual malaise of Delbet was no isolated phenomenon at the turn of the century, and its permeation of the milieu guaranteed an audience for the lucid optimism of Perrin, grounded as it was in an intricate framework of experimental fact and logical ratiocination. Nor was Delbet's the only philosophy of science which adopted the analysis of Perrin in the succeeding years. Émile Meyerson, for example, continually quotes Perrin throughout his *De L'Explication dans les Sciences* (1927), citing not only Perrin's experimental researches, but also utilising his rendition of the historical development of atomistics. Meyerson profiles Perrin's methodology as a prototype of the scientific endeavour.[117]

The name of Perrin had indeed been associated with the philosophy of science since the publication of *Les Principes* and had been incorporated into historical analyses of both physics and its philosophy since Lucien Poincaré's *La Physique Moderne* (1906)[118] and Abel Rey's *La Philosophie Moderne* (1908),[119] with both authors using extensively Perrin's definitions of the first and second laws of thermodynamics and other conceptions. The impact of Perrin's arguments for atoms was certainly no less solid, nor the proliferation of his ideas any less extensive than for thermodynamics. Incorporated into the journals, the philosophy and the history of science, his conception of atomic-molecular reality, as so beautifully promulgated in *Les Atomes*, became exemplary of man's knowledge of matter and of his method of discovering it. The book, perhaps the pinnacle of an epoch of discovery, became a scientific classic. As Jules Sageret wrote in a brief history of the old energetics physics, 'The atomic theory has finally been shown victorious by proof from the most varied experimental verifications. M. Jean Perrin has related them in his book *Les Atomes*, a work truly first-rate . . .'[120]

And What of the Sceptics?

One cannot say, however, that virtually all of the old opponents, without exception, were converted to the new atomistic sect, and indeed, it would seem too much to ask of human nature that this be so. Not only after the conferences and publications of 1908–1913, but even after the more general assimilation of kinetic theory, probability and the molecularity of matter into the corpus of traditional scientific theory and philosophy, there remained a few sceptics on the periphery of the established mentality. Among these were in France, Duhem, Le Chatelier, and Perrin's friend Georges Urbain.

In 1913 in an article in the *Revue Scientifique*, Duhem expressed his opinion of the recent researches on the discontinuity of matter and of the confident assertions that the real sub-structure of matter was now finally being viewed. 'We admit that physical theory can attain a certain knowledge of the nature of things', he partially conceded; but, adhering firmly to his previous position, he maintained that 'this purely analogical knowledge appears to us the limit of progress of theory, the limit which it nears continuously without ever attaining'.[121] The atomistic position he still slapped with the appellation 'une connaissance hypothétique de la nature des choses',[122] and, derogating the new electron or atom of electricity, he ridiculed that 'they' think 'that these hypotheses finally get to the intimate structure of matter, that they are going to show us the elements as if some extraordinary ultramicroscope enlarged them until they are made perceptible to us'.[123] The time will come, he warned, when these models, so dear to the English, will be discarded.[124] Brownian movement he apparently failed to mention explicitly.

Even by the 1920's when Le Chatelier was professor of chemistry at the Sorbonne, he, too, voiced scepticism about the events of the last decades and their definition of molecular reality. When Jean Perrin's son, Francis, enrolled in a course under Le Chatelier, the venerable pedagogue still insisted that the helium which was identified with radioactivity arose, in all probability, from helium impurities in a sample.[125] As mentioned earlier, Georges Urbain, also a professor of chemistry at the Sorbonne and a close friend of Jean Perrin, never accepted these general ideas.[126] This is not to say that he did not find the atomic-molecular theory useful, however, especially in chemistry, but yet he did not view it as representative of an absolute reality, perhaps (as Duhem) extrapolating too far the valid realisation that the contemporary conception was not a final one, that it was not completed in all its possible intricacies.

'The atomistic doctrines are adaptable and flexible: they evolve even as science itself does', Urbain wrote in 1909.[127] Though the fertility of the atomistic approach could not be convincingly disputed, Urbain still hesitated before the yawning pitfall of pure speculation which he feared the approach poorly concealed. The methods of traditional thermodynamics, he also partially distrusted, and as a resolution to his own discomfiture in the face of both physical methods, Urbain could only seek refuge in the remark of Ernst Mach, that 'truth and error have the same sources: it is only success which allows us to distinguish between them'. Urbain added 'That which this thinker understands by "success" is response to experimentation.'[128]

The response of Mach himself to the new proofs for molecular reality proved even more complex than that of Urbain; it was intertwined with the development after 1900 of the opinions of Max Planck, who, it may be remembered, had been a fervent admirer of Mach and anti-atomism at the *fin-de-siècle* and who, by 1909, was publicly attacking Mach with vehement hostility. Mach's reaction was also affected by the publications of Einstein and the correspondence between himself and Einstein. There is no evidence that Mach ever personally met or corresponded with Jean Perrin.

Indeed, Max Planck's attitude towards the statistical, probabilistic interpretation of the second law of thermodynamics had begun to change once he became immersed in the problems of radiation theory, and it was to Boltzmann's methods that he turned in formulating quantum theory,[129] so that by 1910 he was an unwilling, but also a totally convinced, adherent to the kinetic-molecular theory and the probabilistic theory of entropy.[130] In 1908, in a lecture at Leyden, he openly condemned Mach's philosophy as that of a 'false prophet',[131] and revealed that he had wandered from Mach's emphasis upon sense perceptions to the portals of belief in an objectively real physical world, a world which might eventually be portrayed in a unified physical picture if science could only eliminate from its domain the anthropomorphic elements of sense perception.[132] Though it would be totally misleading and erroneous to assert that Planck derived this viewpoint from Jean Perrin, it is accurate to say that Planck's emphasis was not incompatible with that of Perrin and was an (independent) echo of Perrin's lifelong contempt for the strictures of 'puerile anthropomorphism'.[133] Planck, too, now insisted upon the reality of atoms and the validity of their measurements regardless of whether one could actually see them or place them in a measuring pan.[134]

Mach, on the other hand, remained adamant in his censure of the more and more popular conception of the reality of molecules. Indicative of his hostility and its rationale was the unaltered remark in the new 1909 reprint of his essay on the principle of the conservation of work: 'If, then we are astonished at the discovery that heat is motion, we are astonished at something that has never been discovered. It is quite irrelevant for scientific purposes whether we think of heat as a substance or not.'[135] After receiving a copy of the essay, Einstein wrote to Mach that he was sending to Mach some of his own treatises; especially, Einstein requested that Mach glance at the paper on Brownian movement, since here there existed a motion that one must interpret as 'heat motion'.[136] In a postcard which Mach received eight days later from Einstein, Einstein thanked him for a treatise and letter sent from Mach in the meantime: 'I am very happy that you had pleasure in the relativity theory', he writes.[137] So one assumes that Mach had failed to comment on the Brownian movements.

Following this response, or lack of response, were Mach's comments in an article in 1910, where he allowed that the experimental conclusions derived from the atomic theory were most numerous and most useful, but that this proved nothing of the reality of atoms.[138] Further, he warned once again that the world's reality cannot be constructed on physics alone, but must also be founded upon biology.[139] Of the contemporary popular support for the atomic-molecular hypothesis, Mach replied that if physicists were on their way to becoming a church with the essential credo of the reality of the atom, he would renounce the physical way of thinking, as freedom of thought was quite precious to him.[140] Later, in the preface to his *Physical Optics* in July of 1913, Mach wrote, 'But I must as assuredly disdain to be a forerunner of the relativists as I withhold from the atomistic belief of the present day.'[141]

Mach died soon afterwards, apparently then without ever relenting on the question of molecular reality. In later analysing his own motivations in initiating work on the theory of Brownian movement and in explaining the reaction of sceptics such as Mach, Einstein wrote the following:

> Not acquainted with the earlier investigations of Boltzmann and Gibbs, . . . I developed the statistical mechanics and the molecular-kinetic theory of thermodynamics. . . . My major aim in this was to find facts which would guarantee as much as possible the existence of atoms of

> definite finite size. In the midst of this I discovered that, according to atomistic theory, there would have to be a movement of suspended microscopic particles open to observation.

And in the following years these predictions were confirmed:

> The agreement of these considerations with [Perrin's] experience together with Planck's determination of the true molecular size from the law of radiation (for high temperatures) convinced the sceptics, who were quite numerous at that time (Ostwald, Mach) of the reality of atoms. The antipathy of these scholars towards atomic theory can indubitably be traced back to their positivistic philosophical attitude. This is an interesting example of the fact that even scholars of audacious spirit and fine instinct can be obstructed by the interpretation of facts by philosophic prejudices . . . [consisting] in the faith that facts by themselves can and should yield scientific knowledge without free conceptual construction.[142]

Mach not withstanding, Robert Millikan could write in 1917 with perfect confidence and credibility, 'Today there is absolutely no philosophy in the field [of physics] other than the atomic philosophy, at least among physicists.'[143]

Conclusion

The role of Jean Perrin in the definition of the atomic-molecular hypothesis as a physical hypothesis capable of experimental demonstration and his successful verification of that hypothesis by his numerous experiments, most notably those on Brownian movement, is irrefutably central to the resolution of these controversies of the nineteenth century and to the development of the chemistry and physics of the twentieth century. As Stephen Brush has hinted in a recent history, Perrin 'is generally credited with having established the Einstein–Smoluchowski theory by his experiments, but just as important was his role as propagandist for atomism',[144] and indeed, Perrin's own scientific-philosophic posture, which straddled the two methods which he termed 'analogies' and 'intuition', tailored him for the role of arbiter in the polemics which originated between two quite hostile camps.

Jean Perrin stood, in fact, almost midway between positivism and idealism, both in a strictly scientific context and in the body politic at large. His understanding of a probability within determinism, his faith in an evolution which man guides, his belief in individual responsibility and his combination of science and socialism isolated him from both extremes of Right and Left in French politics, tailoring him for a role in Léon Blum's cabinet. Similarly, while his stance in science was clearly mechanistically and materialistically oriented, his philosophical position still insisted upon a precision in definitions and a clarity in logic fully comparable to that in mathematics. Perrin insisted upon the appropriateness of imagination and the usefulness of invisible entities in a deductive method of science and yet did not deny the simultaneous utility of the experiential, inductive approach. His was a position of compromise and of constructive synthesis.

Even the stereotypes of the French versus the English-American and German minds, a nineteenth-century tradition espoused most strongly by Duhem, failed to describe Perrin. One of his students, in later attempting to account for the pivotal role of Perrin in the controversies over thermodynamics and atomistics, ventured the following analysis:

> There are two sorts of physicists. Those of the first kind have sustained the power of latin culture. They guard it with a jealous care from all that which is not strict experimentation. . . . They move forward a little prudently, but assuredly; they are careful to raise a foot only if the other has found the rock. They produce little, but well. They are dominant in France numerically.

The other school of physicists, he continued, was wrapped up in the superficially contradictory concerns of industrialism and mysticism:

> America and Germany are the more important foyers for this. . . . It creates a sort of 'scientific mystique', assigning to the mind a mission of control comparable to the year's-end balance sheets of industrial enterprises. The characteristics here are: intense work, prodigiously unpolished production, important fragments, distinctly large production. On the whole, a gain for society.[145]

Jean Perrin, concluded Marcel Courtines, formed the *trait d'union* of these two schools.[146]

Perrin, then, besides being a scientist of imaginative insight and experimental ingenuity in his diverse researches on cathode rays, X-rays, electrisation of contact, Brownian movement, thin films and later investigations, was a figure of compromise between differing viewpoints and traditions. This is demonstrated not only in his popularised articles, but also in his highly influential texts *Les Principes* and *Les*

Atomes. In accounting for his success as a proponent of the discontinuous hypothesis, one must recall, too, his clarity of exposition, his logic in structure, what Ramsay perhaps too simply and succinctly termed the 'style français'. Yet, even more than this, one must remember the statement of another Frenchman and experimentalist, Claude Bernard, who felt that 'Each great man belongs to his time and can come only at his proper moment, in the sense that there is a necessary and ordered sequence in the appearance of scientific discoveries.'[147]

Indeed, Jean Perrin stepped into the laboratories of chemistry and physics at that hour when investigations had commenced and theory was being mapped out in the three apparently unrelated areas of cathode rays, X-rays and radioactivity; black-body radiation and specific heats; and studies in random process, specifically Brownian movement. Of these, Brownian movement had the longest history and yet was the last problem to be solved. Its resolution, though, followed directly on the heels of the scintillating charged particles and the enigmatic quanta of energy and action. Most certainly, it was both the surprisingly precise correlation of identical parameters—such as the magical number *N*—from superficially unrelated sources, as well as the provision of an inchoate, ultimate sub-structure still eluding the observers of the atom and molecule, which effected the almost universal and uncritical acceptance of the carefully conceived and documented manifestos of Perrin. For the more fortunate *philosophes* among his audience, not only did troubled doubts over the reality of matter disappear, but the dolorous malaise of the 1890s also vanished.

'There is scarcely a physicist or physical chemist of my generation or the following generation who has not been the student of Perrin and Langevin', avowed Edmond Bauer of the Sorbonne, in 1948. 'Three chairs at the Collège de France, others at the Sorbonne, the leadership of the École de physique et de Chimie are occupied by their immediate disciples.'[148] Yet, in addition to these students who actually listened to the exuberant *professeur*, inebriated with excitement over each new interrelationship of molecules and physical phenomena which was documented in these and the next years, there were innumerable scientific colleagues, popular audiences, and university students educated in textbooks based upon the new atomistics or even reading in the original Perrin's own *Les Atomes*.

It can be contended that Perrin's experiments were really the last ostensible victory for the mechanists; that afterwards, relativity, quantum theory, the behaviour of the electron and other perplexing aspects of a new physics produced fundamental physical concepts which are really structures of pure thought, incapable of realisation in any sense which would properly be termed material.[149] If this be true, however, it does not detract from the logic and ingenuity of Perrin's experimental researches in proof of molecular reality nor from his compilation of the multitude of verifications which had been completed by 1913, and which remain the foundation for an understanding of physical phenomena. In addition, the achievements of his scientific career and his lifelong concern with experimental verification for the reality of molecules and the discontinuous structure of matter—a hypothesis of both philosophy and science since the era of the Greeks—worked irrevocably as a catalyst for an immense cascade of succeeding discoveries. A final evaluation of the contributions of Jean Perrin amidst the successful assimilation of molecular reality into the annals of chemistry and physics lies in the commendation of Ernest Renan that 'The highest reward for the scientist is to create a scientific movement by which he himself is surpassed.'[150]

NOTES

1. Stephen Brush, 'A History of Random Processes. I. Brownian Movement from Brown to Perrin', *Archives for the History of the Exact Sciences*, **5** (1968), p. 23.
2. Letter from H. A. Lorentz to Jean Perrin, dated from Leiden, 14 January 1910.
3. Letter from H. Kamerlingh Onnes to Jean Perrin, dated from Leiden, 19 January 1910.
4. Letters from H. Rubens to Jean Perrin, dated from Berlin and the Institut, 12/10/10 and 24/10/10. Rubens congratulates Perrin on his election.
5. Letter from J. H. Poynting to Jean Perrin, dated 20 February 1910.
6. In a letter from E. Branly to Jean Perrin dated from Paris, 30 January 1909, Branly writes of '... the application, so curious, which you have made of the kinetic theory to emulsions of gamboge. Like me, the public has been enchanted.'
7. Letter from Émile Picard to Jean Perrin, dated 30 November 1909.
8. Letter to Jean Perrin dated from Utrecht, 23 September 1909. The author of the letter was professor of physiology at the state veterinary school at Utrecht and wrote that he and Professor H. C. Zwaardemaker had carried out these experiments, which were presented at the ninth *Congrès international de médecine vétérinaire* at Schevennique. The writer further remarked that he had been directed to Perrin's researches by a discourse of Lorentz and that 'I hope it will please you to know that your ideas have found in Holland such a warm welcome from one of our most eminent physicists.' The signature of the writer was most

difficult to decipher, although his name could be traced through the bulletin of the *Congrès*.

9. Letter from W. N. Keesom to Jean Perrin dated from Leyden, 29 November 1911. Keesom writes that his value $s = 0{\cdot}00079$ in the *Annalen der Physik*, **4/35** (1911), p. 590, note 2, was calculated with $N = 70{\cdot}5 \times 10^{22}$ from Perrin's article in the *Physikalische Zeitschrift*, **11** (1910), p. 466.
10. In his memoir of 1909, Perrin included a historical discussion of Brownian movement which mentioned only Wiener, Delsaulx and Carbonnelle, Thirion, and Gouy for their experimental studies.
11. In a letter from William Ramsay to Jean Perrin, dated from 19 Chester Terrace, Regent's Park, N.W., 4 March 1912 (written in French), Ramsay writes of his own work, that in an 1876 memoir in the *Proceedings of the Geological Society*, 'I attribute the movement to the influence of heat, which causes molecular movement. It is nearly the same reasoning that M. Gouy has employed.' Ramsay also directed Perrin to Jevon's brochure on Brownian movement, and suggested that Buffon had identified the Brownian particles with his living molecules ('Buffon believed that it was life') and that Spallanzani had named them 'animaletti d'ultimo ordine'.
12. In a letter from Gouy to Perrin, dated from Lyon, 31 October 1908, Gouy writes to thank Perrin for his friendly letter and 'l'envoi qui l'accompagnait'.
13. In another letter from Gouy to Perrin, dated from Lyon, 18 Nov., Gouy writes 'I have read and re-read your grand and beautiful memoir, and I send you all my congratulations.'
14. Letter from Gouy to Perrin, dated only 'jeudi' and from Lyon. From the context, the letter was probably written around early 1910.
15. Ibid.
16. The Svedberg, *Die Existenz der Moleküle* (Leipzig: Akademische Verlagsgesellschaft m.b.H., 1912), p. 117.
17. 'Mouvement brownien et réalité moléculaire', in the *Oeuvres Scientifiques*, p. 129.
18. Letter from Smoluchowski to Jean Perrin, dated from Lemberg (Lvov, Poland), 24/1/1909 (and written in French).
19. According to Francis Perrin.
20. Letter from Einstein to Jean Perrin, dated from Zurich, 11/xi/09.
21. Of course, it was Perrin's value which was off, as explained in Chapter III, f. 105.
22. See Chapter III, f. 59.
23. 'Ueber einen Satz der Wahrscheinlichkeitsrechnung und seine Anwendung in der Strahlungstheorie', *Annalen der Physik*, (4) **33** (1910), pp. 1096–1104 and 'Statistische Untersuchung der Bewegung eines Resonators in einem Strahlungsfeld', *Annalen der Physik*, (4), **33** (1910), pp. 1105. Professor Erwin Hiebert directed my attention to the collaboration between Einstein and Hopf. Hopf was not only a scientific *confrère*, but also a close friend of the Einstein family and a professor at the Technische Hochschule in Aachen.
24. Letter from Einstein to Hopf, dated only 'dienstag', but presumably from Zurich and probably in December of 1910 or January of 1911 (because of the place of the letter in the total context of correspondence and published papers, and

because of the final greeting 'Beste Glückswunsche zum Jahrewechsel'). Courtesy of Miss Helen Dukas of the Einstein Collection at the Institute for Advanced Study in Princeton.

25. Ibid.
26. Einstein, 'Berechtigung zu meiner Arbeit: 'Eine neue Bestimmung der Molekül-dimensionen" ', *Annalen der Physik*, **34** (1911), pp. 591–592.
27. Letter from Einstein to Perrin, dated from Zurich, 12/1/11. There remains a slight problem in interpretation, however, as Einstein writes to Hopf that Bancelin's experiments indicated a coefficient of 3·8, while Einstein's theory had utilised 1. Einstein's new result was 2·5, which does indeed vary quite a bit from 3·8, but in 1912, in his article *Les preuves . . . moléculaires*, p. 35, Perrin refers to Bancelin's experimental coefficient as 2·8.
28. Letter from Soddy to Jean Perrin, dated from The University, Glasgow, January 11, 1909.
29. Taylor and Francis: London, 1910; and the *Beihefte*, **1** (1910), p. 221.
30. E. E. Fournier d'Albe, 'Brownian Motions', *English Mechanic and World of Science*, **88**, #2282 (1908), p. 459.
31. D'Albe, 'Theory of Brownian Motions', *English Mechanic and World of Science*, **88**, #2285 (1909), p. 529.
32. Stephen Jahn, 'Über einen Fortschritt in der atomischen Auffassung der Materie', *Jahrbuch der Radioaktivität und Elektronik*, **6** (1909), pp. 229–242.
33. Ibid., p. 242.
34. Werner Mecklenburg, *Die experimentelle Grundlegung der Atomistik* (Jena: Gustav Fischer, 1910). Mecklenburg was an assistant at the Chemisches Institut, University of Jena, from 1907–1908 and then an 'Assist., Material-prüfgsamt, Berlin-Dahlen, 1908–1914' (Poggendorf). He was at the mining academy in Clausthal subsequently.
35. Letter from Werner Mecklenburg to Jean Perrin, dated 29 January 1910. An earlier letter of 13 January 1910 is dated from Clausthal im Harz.
36. Mecklenburg, *Die experimentelle Grundlegung der Atomistik*, p. 134.
37. Florian Cajori, *A History of Physics* (New York; Dover, 1962), p. 226.
38. Nernst, *Theore ische Chemie*, 10th ed. (Stuttgart: Ferdinand Enke, 1921), p. 224. Also quoted in Brush, pp. 34–35.
39. Svante Arrhenius, 'La théorie moléculaire', *Revue du Mois*, **11** (1911), p. 388.
40. Ibid., p. 392.
41. Quoted from Brush, p. 35.
42. Wilhelm Ostwald, 'Büchershau', *Zeitschrift für physikalische Chemie*, **64** (1908), p. 509. Ostwald's detailed statement is the following: 'The writer has himself already expressed on an earlier occasion his opinion that the long-sought experimental proof for the kinetic theory appears to have been treated in this work; and he believes that, with the satisfactory dispatch of these proofs . . . the cause of his rejection of this theory—which even in its lack of such an experimental proof has managed to endure—is now gone. Perhaps he may be permitted to add personally that he views his previous opposition neither as a loss for himself, nor as one for science. For the critical and rationalising task bound up with it [science] has been necessary under all circumstances, and since up until now it [kinetic

theory] had not been explained by the followers of the atomic theory, an opponent was required.'

43. Wilhelm Ostwald, Vorbericht to *Grundriss der physikalischen Chemie* (Grossbothen, November 1908).
44. Ibid.
45. Fragment of a letter from Ostwald to Jean Perrin, dated 25/4/1913.
46. In a letter from Wolfgang Ostwald to Jean Perrin, dated from Leipzig, 14 Oct. 1909, Ostwald requests permission to translate Perrin's 'letzte Abhandlung' into the *Kolloidchemische Beihefte*.
47. Ibid.
48. Ibid. Wolfgang Ostwald wrote: 'I am working now on a book, the title of which will run: Energetic Atomistics. Indeed I am working here from a somewhat different point of view, insofar as I find, for example, that the fact of discontinuity is synonymous with the existence of an immense curved surface, and therefore that surface and volume energies, as they appear intimately in capillary or surface pressure, for example, are of the same general applicability and importance as discontinuity itself.'
49. Eugène Bloch, 'L'atomisme dans la physique contemporaine', *Revue du Mois*, **13** (1912), pp. 146–164, and J. H. Boex-Borel (J. H. Rosny-aîné), 'Les principes de l'énergétique', *Revue du Mois*, **13** (1912), pp. 385–395.
50. Abel Rey, *L'énergétique et le mécanisme* (Paris: Félix Alcan, 1907), p. 80.
51. Letter from Louis Olivier to Jean Perrin, dated 29 September 1908. Olivier was the *directeur* of the *Revue générale des sciences*.
52. *Revue du Mois*, **14** (1912), pp. 513–540.
53. Jean Perrin, 'Mouvement brownien et molécules', *Journal de Physique théorique et appliquée*, **9** (1910), pp. 5–39.
54. Jean Perrin, 'Conférence sur le mouvement brownien' (École professionelle supérieure des Postes et Télégraphes, 1910). 10 pp.
55. Jean Perrin, 'La réalité des molécules,' *Revue Scientifique*, **49** (2) (1911), pp. 774–784.
56. Ibid., p. 774.
57. Ibid., p. 776.
58. Letter from Ernest Rutherford to Jean Perrin, dated from 17 Wilmslow Road, Withington, Manchester, 4 May 1909.
59. Jean Perrin, 'Le mouvement brownien', Conférence à la Royal Institution, Londres, 24 fevrier 1911.
60. Letter from William Crookes to Jean Perrin, dated from 7 Kensington Park Gardens, London, W., 24 October 1910.
61. Letter from F. Kurlbaum to Jean Perrin, dated from Berlin, 25 April 1911.
62. Maurice de Broglie, *Les Premiers Congrès de Physique Solvay et l'orientation de la physique depuis 1911* (Paris, 1951), p. 12.
63. Ibid., pp. 101–102.
64. Ibid., p. 16.
65. Ibid., pp. 15–20.
66. Ibid., pp. 21–22.
67. In ibid., pp. 27–30, M. de Broglie briefly résumés the progress of the Congress.

For the complete reports and discussions of the five-day meetings, see *La théorie du rayonnement et la quanta*, publié par P. Langevin et M. de Broglie (Paris: Gauthier-Villars, 1912).

68. Jean Perrin, 'Les preuves de la réalité moléculaire', pp. 153–250 in *La théorie du rayonnement et les quanta*, p. 164.
69. *La théorie du rayonnement et les quanta*, p. 251. The ensuing discussions following each paper at the Congress are recorded in this volume.
70. Ibid.
71. Jean Perrin, 'Les Preuves de la réalité moléculaire', pp. 1–53 in *Les Idées Modernes sur la Constitution de la Matière* (Paris: Gauthier-Villars, 1913), pp. 23–24.
72. For the complete memoirs and discussions, see *Les Idées Modernes sur la constitution de la matière*.
73. Henri Poincaré, 'The principles of mathematical physics', trans. George B. Halsted, pp. 604–622 in *Congress of Arts and Science. Universal Exposition, St. Louis, 1904*, ed. Howard J. Rogers (New York: Houghton, Mifflin, & Co., 1905), pp. 621–622.
74. Paul Langevin, 'L'Oeuvre d'Henri Poincaré', *Revue de Métaphysique et Morale*, **21** (1913), pp. 675–718.
75. Henri Poincaré, 'Les Rapports de la Matière et de l'Éther', extract from the *Journal de Physique*, mai 1912 (Paris: Deslis Frères, 1912), 16 pp. Also in *Les Idées modernes sur la Constitution de la Matière*.
76. Perrin, 'Les hypothèses moléculaires', *Revue Scientifique*, **15** (1901), p. 451.
77. Jean Perrin, *Les Atomes*, 4th ed. (Paris: Félix Alcan, 1914), pp. 72–73.
78. Ibid., pp. 74–75.
79. Ibid., p. 75.
80. Jean Perrin, 'Observations sur les lames minces', Conférence à la séance annuelle de la Société suisse de Physique, et à la Société de Chimie physique à Paris, avril 1913; 'La stratification des lames liquides', *Annales de Physique*, **10** (1918), p. 160; and 'Pellicules monomoléculaires et lames liquides feuilletées', *Notice sur les Travaux Scientifiques* (Toulouse: Edouard Privat, 1923), pp. 45–59 and in the *Oeuvres Scientifiques*, pp. 248–261.
81. Jean Perrin, 'La structure discontinue de la matière', Conférence Nobel, Stockholm, 12 decembre 1926, pp. 139–165 in the *Nobel Lectures, Physics, 1922–1941* (Amsterdam, 1965), p. 158. Perrin's discourse was also published at Stockholm as a separate brochure (Imprimerie Royale. P. A. Norstedt & Fils, 1927, 29 pp.).
82. Jean Perrin, 'La stratification des lames liquides', pp. 104–122 in Perrin's *Notice sur les Titres et Travaux Scientifiques* (Paris: Gauthier-Villars, 1918), pp. 108–110.
83. Ibid., p. 115.
84. Conférence Nobel, p. 159.
85. *Notice* . . . (1918), p. 120.
86. Conférence Nobel, p. 156.
87. For example, in the second English edition (1923), based on the eleventh French edition, Perrin includes footnotes, p. 50, directing the reader to both his and P. V. Wells' papers of 1918 and 1921, respectively.
88. *Oeuvres Scientifiques*, bibliography, p. xii.
89. *Les Atomes* (1914), p. xi.

90. See Chapter II.
91. *Les Atomes* (1914), p. iv.
92. Ibid., p. 205.
93. In a later revision to this chapter on light and quanta, Perrin augments the applications of quantum theory with several new investigations, including his own research into the role of light in chemical reactions. As was mentioned in Chapter II, these investigations were carried out after the war and in the 1920's, and a number of valuable experimental studies emerged from Perrin's basic interest. His fundamental assumption was that, 'Since ... the rate of dissociation depends largely on the temperature, we are reminded that temperature exerts its influence by radiation as well as through molecular impact, and are faced with the suggestion that the cause of dissociation lies in the visible and invisible *light* that fills, under stationary conditions, the isothermal ... enclosure wherein the molecules are moving. *The essential mechanism of all chemical reaction is therefore to be sought in the action of light upon atoms*' (*Atoms*, trans. D. L. Hammick, 2nd English ed.; New York: Van Nostrand, 1923, p. 164). This assumption was firmly doubted by a majority of physicists, and after reading Perrin's 'Matière et lumière' (*Annales de physique*, **11** (1919), pp. 1–108) upon Perrin's request (letter from Jean Perrin to Albert Einstein, dated from Paris, 28 August 1919), Einstein replied in a letter of 5 November 1919 (written in French), that unfortunately, 'Your opinion of the primary importance of radiation for all chemical reactions still appears doubtful to me ...' Both letters are in the Einstein Collection at the Institute for Advanced Study in Princeton.
94. *Les Atomes* (1914), pp. 293–295.
95. Ibid., p. 293.
96. See note 43.
97. In a note signed by Lucien Poincaré and written on his calling card, dated 4 February 1913.
98. Letter from Jacques Loeb to Jean Perrin, dated from the Rockefeller Institute for Medical Research, Marine Laboratory, Woods Hole, Mass., 7 August 1918.
99. John Bernal, pp. 40–43 in *Hommage National à Paul Langevin et Jean Perrin* (Paris: Orleans, 1948), p. 41.
100. Ibid.
101. Letter, dated 4 March 1912.
102. See Chapter I. Guye was professor of chemistry at the University of Geneva and publisher of the *Journal de chimie physique*.
103. Ph. A. Guye, review of *Les Atomes*, in *Journal de chimie physique*, **11** (1913), p. 169.
104. Letter from Ph. A. Guye to Jean Perrin, dated 1 February 1913.
105. Anonymous review of *Les Atomes*, in *Nature*, **91** (March–August, 1913), p. 473.
106. For example, Guye refers the reader to Svedberg's book for even further information. Aldo Mieli discusses them both in a reprint entitled 'Jean Perrin—Les Atomes. The Svedberg—*Die Existenz der Moleküle*. M. Harten—*Die Hauptlehren des Averroes nach seiner Schrift*. Geza Revesz—*Zur Grundlegung der Tonpsychologie*' (Genova: A. F. Formiggini, 1913).
107. Letter from J. J. Thomson to Jean Perrin, dated from Cambridge, 10 February 1913.

108. Edward W. Washburn, *avant-propos* to *Introduction aux Principes de la Chimie Physique du point de vue de l'atomistique et de la thermodynamique*, trans. Henri Weiss and W. Albert Noyes, pref. by Jean Perrin (Paris: Payot, 1925), p. viii.
109. Ibid.
110. J. Lamirand and Ch. Brunold, *Cours de Chimie pour les Classes de Mathématiques-Speciales* (Paris: Masson et Cie, 1927). Perrin commended the text for its introductory treatise on physical chemistry and atomism.
111. Pierre Delbet, *La Science et la Réalité* (Paris: Flammarion, 1913), pp. 1–2.
112. Ibid., p. 3.
113. Ibid., p. 6.
114. Ibid., p. 186.
115. Ibid., p. 313.
116. Ibid., p. 314.
117. Émile Meyerson, *De L'Explication dans les Sciences* (Paris: Payot, 1927), 784 pp.
118. Paris: Flammarion, 1904, 311 pp.
119. Paris: Flammarion, 1908, 372 pp.
120. Jules Sageret, *La Vague Mystique* (Paris: Flammarion, 1920), p. 180.
121. Pierre Duhem, 'Examen logique de la théorie physique', *Revue scientifique*, **51** (1) (1913), p. 740.
122. Ibid.
123. Ibid.
124. Ibid.
125. As remarked by Francis Perrin.
126. Ibid.
127. Georges Urbain, 'Le Rôle de l'atomistique dans l'enseignement et dans la recherche', *Revue du Mois*, **7** (1909), pp. 417–429. This paper is extracted from the first lesson (2 mars 1909) of the course of chemistry at the Faculty of Sciences of Paris, as M. Urbain replaced Alfred Ditte.
128. Ibid., pp. 428–429.
129. Erwin N. Hiebert, 'The Conception of Thermodynamics in the Scientific Thought of Mach and Planck', *Ernst Mach Institut. Wissenschaftlicher Bericht*, #5 (1968), pp. 104–105.
130. Ibid., p. 89.
131. Ibid., p. 78 Quoted from p. 29 of Max Planck, 'Die Einheit des physikalischen Weltbildes' (1909) in Planck's *Abhandlungen und Vorträge*.
132. Ibid., p. 73.
133. See Chapter II, f. 60.
134. Weinberg, *Mach's Empirio-Pragmatism in Physical Science* (New York: Albec Press, 1937), p. 54.
135. Quoted in Brush, 'A History of Random Process. I. Brownian Movement from Brown to Perrin,' p. 36, from the English translation (Chicago, 1911), p. 37. Also quoted in Hiebert, p. 3.
136. Quoted in Hiebert, pp. 78–79, from a letter from Einstein to Mach, dated from Bern, 9 August 1909, published by Friedrich Herneck, courtesy of Frau Annai Karma Mach: 'Zum Briefwechsel A lbert Einsteins mit Ernst Mach (Mit zwe unveröffentlichten Einstein-Briefen)', *Forschungen und Fortschritte*, **37** (1963).

pp. 239–243. Einstein wrote (my translation), 'Because I do not know how else to show you my thanks, I am sending you several of my papers. I would especially like to ask you briefly to take a look at the one concerning Brownian movement, because here lies a motion which must be interpreted as "heat movement".'

137. Quoted in Hiebert, p. 79, from a card from Einstein to Mach, dated from Bern, 17 August 1909, published by Friedrich Herneck, courtesy of Frau Anna Karma Mach, 'Nochmals über Einstein und Mach', *Physikalische Blätter*, **17** (1961), pp. 275–276.
138. Hiebert, p. 84, from Ernst Mach, 'Die Leitgedanken meiner naturwissenschaftlichen Erkenntnislehre und ihre Aufnahme durch die Zeitgenossen', *Physikalische Zeitschrift*, **11** (1910), pp. 599–606. Also in *Scientia*, **7** (1910), pp. 225–240.
139. Hiebert, p. 84.
140. Ibid., p. 85. He wrote in answer to Planck's accusation of 'false prophet' (my translation) 'One sees that physicists are on the direct way to becoming a church. . . . If belief in the reality of the atom is so essential for them, then I myself renounce the physicists' way of thinking; then I do not want to be a correct physicist; I renounce every scientific value; in short, I thank you very much for the community of belief. For freedom of thought is more dear to me.'
141. Ernst Mach, *The Principles of Physical Optics*, trans. John Anderson and A. F. A. Young (New York: Dover, 1926), p. viii. The original German reads: 'Ich muss es aber nun mit der selben Entschiedenheit ablehnen, den Relativisten vorangestellt zu werden, mit welcher ich die atomistische Glaubenslehre der heutigen Schule oder Kirche für meine Person abgelehnt habe' (*Die Prinzipien der physikalischen Optik*; Leipzig: J. A. Barth, 1921; p. viii).
142. Albert Einstein, 'Autobiographical Notes', pp. 1–95 in *Albert Einstein: Philosopher-Scientist*, ed. Paul Arthur Schilpp (New York: Harper Torchbooks, 1959), pp. 47–48.
143. Robert Andrews Millikan, *The Electron* (Chicago: University of Chicago Press, 1917), p. 10.
144. Brush, p. 30.
145. Marcel Courtines, 'La lumière principe du monde à propos de Jean Perrin', *Cahiers de la Quinzaine*, 18e serie, **4** (Paris: L'Artisan du Livre, 1927), pp. 8–9.
146. Ibid., p. 10.
147. Claude Bernard, *An Introduction to the Study of Experimental Medicine*, trans. Henry Copley Green (New York: Dover, 1957), p. 41.
148. Edmond Bauer, pp. 20–23 in *Hommage National à Paul Langevin et Jean Perrin*, p. 22.
149. Philipp Frank, *Modern Science and its Philosophy* (Cambridge: Harvard University Press, 1949), p. 187, on the basis of James Jeans' conception of physics in *The Mysterious Universe*.
150. Ernest Eslangon, 'Des membres et des correspondants de l académie', *Comptes rendus*, **214** (1942), p. 729, quoting Renan.

Bibliography

In analysing the role of Jean Perrin in the controversies over molecular reality and the discontinuous hypothesis, several types of bibliographical material have been used. These have been grouped below simply as published and unpublished sources. An almost complete bibliography of Perrin's publications may be found in the *Oeuvres Scientifiques de Jean Perrin* published by the Centre National de la Recherche Scientifique in 1950 (see pp. vii-xii).

The unpublished sources were largely made available to me through the kindness of Professor Francis Perrin who allowed me, especially, the use of a number of letters in his keeping, and who most graciously spent an evening answering many unsolved questions which were on my mind. Professor Erwin N. Hiebert also obtained for my use photocopies of several letters in the Einstein Collection at the Institute for Advanced Study at Princeton.

No terse abbreviations are used for the journals listed in the bibliography, although the *Comptes rendus de l'Académie des Sciences* are referred to only as *Comptes rendus*. Both the *Monthly Microscopical Society Journal* and the *Royal Microscopical Society Journal* are cited, although the former was renamed as the latter around 1880. In the discussions of Jean Perrin's published papers in the text of this analysis, page numbers in the *Oeuvres Scientifiques* are often cited, while these are not given in the bibliography. Those papers which may be found in whole or in part in the *Oeuvres Scientifiques* are simply starred.

Published Sources

Alexander, Peter. 'The Philosophy of Science, 1850–1910', pp. 402–425 in D. J. O'Connor (ed.), *A Critical History of Western Philosophy*. London: Collier-Macmillan, 1964.

Arrhenius, Svante. 'La théorie moléculaire', *Revue du Mois*, **11** (1911), 358–398.

Atomes et Électrons. Institut Internationale de Physique Solvay à Bruxelles, 1–6 avril 1921. Paris: Gauthier-Villars, 1923.

Bache, R. Meade. 'The Secret of the Brownian Movements', *Proceedings of the American Philosophical Society*, **33** (1894), 163–177.

Bachelier, Louis. Mémoires, *Annales de l'École normale supérieure*, **17** (1900), 21–86; **18** (1901), 143–210.

Becker, Carl L. *The Heavenly City of the Eighteenth-Century Philosophers*. New Haven: Yale University Press, 1932.

Bernard, Claude. *An Introduction to the Study of Experimental Medicine*. Trans. Henry Copley Green. New York: Dover, 1957.

Berthelot, Marcelin. *Les Origines de l'Alchimie*. Paris: Georges Steinheil, 1885.

Blum, Léon. 'Jean Perrin', pp. xxix–xxxv in Jean Perrin, *La Science et l'Espérance*. Paris: Presses Univérsitaires de France, 1948.

Bloch, Eugene. 'L'atomisme dans la physique contemporaine', *Revue du Mois*, **13** (1912), 146–164.

Boltzmann, Ludwig. *Populäre Schriften*. Leipzig: J. A. Barth, 1905.

Vorlesungen über die Principe der Mechanik. 2 vols. Leipzig: J. A. Barth, 1897–1904.

Borel, Émile. 'Transfert des cendres de Jean Perrin', *Institut de France. Académie des Sciences*, #11 (1948), 5 pp.

Boussinesq, Joseph. 'Recherches sur les principes de la mécanique, sur la constitution moléculaire des corps et sur une novelle théorie des gaz parfaits', *Memoires de l'Académie des Sciences et Lettres de Montpellier*, 8 juillet 1872, 48 pp.

Brevoort, H. L. 'Brownian movement in milk', *New York Microscopical Society Journal*, **2** (1886), 101.

Brewster, David. 'Observations relative to the motions of the molecules of bodies', *Edinburgh Journal of Science*, **10** (1829), 215–220.

Brillouin, Marcel. Introduction and preface to Ludwig Boltzmann, *Leçons sur la théorie des gaz*, trans. A. Gallotti. 2 vols. Paris: Gauthier-Villars, 1902–1903.

'Pour la Matière', *Revue générale des sciences*, **6** (1895), 1032–1034.

Recherches récentes sur diverses questions d'hydrodynamique. Paris: Gauthier-Villars, 1891.

Brock, W. H. and D. M. Knight. *The Atomic Debates*. Leicester University Press, 1967.

Broglie, Louis de. 'Jean Perrin', pp. ix–xxviii in *La Science et l'Espérance*.

'La réalité de molécules et l'oeuvre de Jean Perrin', *Institut de France. Académie des Sciences Mémoires*, ser. 2, **67** (1949), 29 pp.

New Perspectives in Physics. Trans. A. J. Pomerans. Edinburgh, 1962.

Francis Perrin and Jean Coulomb. 'Hommage national à Jean Perrin à l'occasion du vingtième anniversaire de sa mort à Paris', *Institut de France. Académie des Sciences*, #14 (1962), 27 pp.

Broglie, Maurice de. *Les Premiers Congrès de Physique Solvay et l'Orientation de la Physique depuis 1911*. Paris, 1951.

Brongniart, Adolphe. 'Mémoire sur la génération de la développement de l'embryon dans les végétaux phanerogames', *Annales des Sciences Naturelles*, **12** (1827), 14–23, 145–196, 255–296.

'Nouvelles recherches sur le pollen et les granules spermatiques des végétaux', *Annales des Sciences Naturelles*, **15** (1828), 381–401.

Brown, Robert. 'A brief account of microscopical observations made in the months of June, July, and August, 1827, on the particles contained in the pollen of plants; and on the general existence of active molecules in organic and inorganic bodies', *Philosophical Magazine*, **4** (1828), 161–173.

'Additional remarks on active molecules', *Philosophical Magazine*, **6** (1829), 161–166.

'Brownian Movement', *Royal Microscopical Society Journal*, **10** (1890), 120. This article discusses an 1828 letter from Rev. Henry to Rev. Buckland inserted in the *Phil. Mag.* issue containing Brown's article.

'Brownian Movement', *Royal Microscopical Society Journal*, **15** (1895), 591. A summary of M. C. Maltezos' article in *Comptes rendus* (1895).

'Brownian or pedetic motion', *Royal Microscopical Society Journal*, **3** (1883), 611–612. A summary of William Ramsay's article on Brownian movement in the *Proceedings of the Bristol Naturalists' Society* (1882).

Brush, Stephen G. 'A History of Random Process. I. Brownian Movement from Brown to Perrin', *Archives for the History of Exact Sciences*, **5**, #1 (1968), 1–36.

Budde. 'Untersuchungen über die Brownsche Molekülarbewegung', *Verhandlungen des Naturhistorischen Vereines der preussischen Rheinlande und Westphalens*, **27** (1870), 108–110.

Burton, E. F. 'The action of electrolytes on colloidal solutions', *Philosophical Magazine*, **12** (1906), 472–478.

'On the properties of electrically prepared colloidal solutions', *Philosophical Magazine*, **11** (1906), 425–447.

Cabannes, Jean. 'Transfert des cendres de Jean Perrin', *Institut de France. Académie des Sciences*, #10 (1948), 5 pp.

Cajori, Florian. *A History of Physics*. New York: Dover, 1962.

Cannizzaro, Stanislao. 'Considerations on some points of the theoretic teachings of chemistry', pp. 17–33 in C. S. Gibson and A. J. Greenaway (eds.), *Faraday Lectures, 1869–1928*. London: The Chemical Society, Burlington House, 1928.

Cardwell, D. S. L. (ed.). *John Dalton and the Progress of Science*. Manchester University Press, 1968.

Caullery, Maurice. *La science française depuis le XVIIe siècle*. Paris: Armand Colin, 1948.

Chaudesaigues. 'Le mouvement brownien et le formule d'Einstein', *Comptes rendus*, **147** (1908), 1044–1046.

Cohen, Robert S. 'Hertz's Philosophy of Science: An Introductory Essay', in Heinrich Hertz, *Principes of Mechanics*, trans. D. E. Jones and J. T. Walley. New York: Dover, 1956.

Conn, G. K. T. and H. D. Turner. *The Evolution of the Nuclear Atom*. New York: American Elsevier, 1965.

Cornu. 'Quelques mots de réponse à "La déroute de l'atomisme contemporain" ', *Revue Générale des Sciences*, **6** (1895), 1030–1031.

Cotton, Aimé. 'Recherches récentes sur les mouvements browniens', *Revue du Mois*, **5** (1908), 737–741.

and Henri Mouton, 'Étude directe du transport dans la courant des particules ultramicroscopiques', *Comptes rendus*, **138** (1904), 1584–1586.

'Transport dans le courant des particules ultramicroscopiques', *Comptes rendus*, **138** (1904), 1692–1699.

Les Ultramicroscopes et les Objets Ultramicroscopiques. Paris: Librairie de l'Académie de Médicine, 1906.

Courtines, Marcel. 'La lumière principe du monde. À propos de Jean Perrin, prix Nobel de physique 1926', *Cahiers de la Quinzaine*, (18), **4**. Paris: L'Artisan du Livre, 1927. 63 pp.

Crookes, William. 'Les théories modernes sur la matière', *Revue Scientifique*, (4), **20** (1903), 225–233.

Dancer, J. B. 'Remarks on Molecular Activity as shown under the microscope', *Proceedings of the Manchester Literary and Philosophical Society*, **7** (1868), 162–165.

Dantzig, Tobias. *Henri Poincaré, Critic of Crisis*. New York: Charles Scribner's Sons, 1954.

Davy, Humphry. 'Discourses of the President, 1826', pp. 92–99 in Vol. 7 of the *Collected Works of Sir Humphry Davy* (ed.) John Davy. London: Smith, Elder, and Co., 1840.

Delbet, Pierre. *La Science et la Réalité*. Paris: Flammarion, 1913.

Delsaulx, Joseph. 'Thermodynamic Origin of the Brownian Motions', *Monthly Microscopical Journal*, **18** (1877), 1–7.

'Deux savants français: Jean Perrin et Paul Langevin', Paris: Georges Lang, 1948. 16 pp.

Donnan, F. G. 'Ostwald Memorial Lecture', *Journal of the Chemical Society* [London], Part I (1933), 316–322.

Duclaux, Jacques. 'Pression osmotique et mouvement brownien', *Comptes rendus*, **147** (1908), 131–134.

'Recherches sur les substances colloïdales', *Journal de chimie physique*, **5** (1907), 29–56.

Dugas, René. *La Théorie Physique au Sens de Boltzmann et ses Prolongements Modernes*. Paris: Dunod, 1959.

Duhem, Pierre. 'Examen logique de la théorie physique', *Revue Scientifique*, **51** (1913), 737–740.

La système du monde, histoire des doctrines cosmogoniques de Platon à Copernic, Vol. I. Paris: Hermann, 1913.

The Aim and Structure of Physical Theory. Trans. from the 2nd Fr. ed., Philip Wiener. Princeton: Princeton University Press, 1954.

Traité d'énergétique. 2 vols. Paris: Gauthier-Villars, 1911.

Dujardin, Félix. *Nouveau Manuel Complet de l'Observateur au Microscope*. Paris: Roret, 1843.

Ehrenhaft, F. 'Die Brownsche Molekularbewegung in Gasen', *Sitzungsanzeiger der kaiserlicher Akademie der Wissenschaft in Wien, mathem.-naturw. Klasse*, **4** (February, 1907), 72.

Einstein, Albert. 'Autobiographical Notes', pp. 1–95 in Vol. I of Paul Arthur Schilpp (ed.), *Albert Einstein: Philosopher-Scientist*. 2 vols. New York: Harper Torchbooks, 1959.

'Berechtigung zu meiner Arbeit: "Eine neue Bestimmung der Molekül-dimensionen" ', *Annalen der Physik*, **34** (1911), 591–592.

'Eine neue Bestimmung der Molekül-dimensionen', *Annalen der Physik*, **19** (1906), 289–306.

'Elementare Theorie der Brownschen Bewegung', *Zeitschrift für Electrochemie*, **14** (1908), 235–239.

Investigations on the Theory of the Brownian Movement (ed.) R. Fürth. Trans. A. D. Cowper. New York: Dover, 1956.

'Theoretische Bemerkungen über die Brownsche Bewegung', *Zeitschrift für Electrochemie*, **13** (1907), 41–42.

'Theorie der Brownschen Bewegung', *Annalen der Physik*, **19** (1906), 371–381.

'Über die von der molekular-kinetischen Theorie der Wärme Bewegungen von in ruhenden Flüssigkeiten suspendierten Teilchen', *Annalen der Physik*, **17** (1905), 549–560.

Eslangon, Ernest. 'Jean Perrin. Des membres et des correspondants de l'académie', *Comptes rendus*, **214** (1942), 725–729.

Exner, Felix. 'Notiz au Brown's Molekularbewegung', *Annalen der Physik*, **2** (1900), 843–847.

Exner, Sigmund. 'Untersuchungen über Brown's Molekularbewegung', *Sitzungsberichte der kaiserlichen Akademie der Wissenschaften, mathematisch-naturwissenschaftliche Klasse, Wien*, **56** (1867), 116–123.

Fournier d'Albe, E. E. 'Brownian Motions', *English Mechanic and the World of Science*, **88**, #2282 (1908), 459–460.

'Theory of Brownian Motion', *English Mechanic and the World of Science*, **88**, #2285 (1909), 529–530.

Frank, Philipp. *Modern Science and Its Philosophy*. Cambridge: Harvard University Press, 1949.

Freund, Ida. *The Study of Chemical Composition.* Cambridge: Cambridge University Press, 1904.

Gillispie, Charles Coulston. *The Edge of Objectivity.* Princeton: Princeton University Press, 1960.

Gouy, Léon. 'Le Mouvement brownien et les mouvements moléculaires', Lyon: A. Storck, 1895. 23 pp. Originally published in the *Revue Générale des Sciences,* **6** (1895).

'Note sur le mouvement brownien', *Journal de Physique,* **7** (1888), 561–564.

'Sur le mouvement brownien', *Comptes rendus,* **109** (1889), 102–105.

Gray, Stephen. 'Several microscopical observations and experiments', *Philosophical Transactions,* **19** (1696), 280–287.

Guye, Phillipe A. Review of Jean Perrin's *Les Atomes. Journal de chimie physique,* **11** (1913), 169.

Haas-Lorentz, G. L. de. *Die Brownsche Bewegung und einige verwandte Erscheinungen.* Friedv. Vieweg, 1913.

Haller, M. A. 'Inauguration de la statue de Adolphe Wurtz'. Institut de France. Académie des Sciences. Paris: Gauthier-Villars, 1921. 11 pp.

Hannequin, Arthur. *Essai critique sur l'hypothèse des atomes dans la science contemporaine.* 2nd ed. Paris: Alcan, 1899.

Hartley, Walter N. 'An Explanation of the "Brownian Movement" ', *Monthly Microscopical Society Journal,* **18** (1877), 8–9.

'On Attraction and Repulsion of Bubbles by Heat', *Proceedings of the Royal Society,* **26** (1877), 137–149.

Henri, Victor. 'Étude cinématographique des mouvement brownien', *Comptes rendus,* **146** (1908), 1024-1026.

'Influence du milieu sur les mouvements browniens', *Comptes rendus,* **147** (1908), 62–65.

Hertz, Heinrich. *The Principles of Mechanics.* Trans. D. E. Jones and J. T. Walley. New York: Dover 1956.

Hiebert, Erwin N. 'The Conception of Thermodynamics in the Scientific Thought of Mach and Planck', *Ernst Mach Institut. Wissenschaftlicher Bericht,* **5** (1968), 106 pp.

Ihde, Aaron J. *The Development of Modern Chemistry.* New York: Harper and Row, 1964.

Infeld, Leopold. 'On the Theory of Brownian Movement', *University of Toronto Studies. Applied Mathematics Series,* #4 (1940), 42 pp.

Jahn, Stephen. 'Über einen Fortschritt in der atomischen Auffassung der Matiere', *Jahrbuch der Radioaktivität und Electronik,* **6** (1909), 229–242.

Jevons, William Stanley. 'On the so-called Molecular Movements of Microscopic Particles', *Proceedings of the Manchester Literary and Philosophical Society,* **9** (1869–1870), 78–84.

Jubilé de M. Marcel Brillouin. Mémoires originaux offerts à M. Marcel Brillouin à l'occasion de son 18e anniversaire. Paris: Gauthier-Villars, 1935.

Kockelmans, J. *Philosophy of Science. The Historical Background.* New York: The Free Press, 1968.

Körber, Hans-Günther, and Grete Ostwald (eds.). *Aus dem Wissenschaftlicher Briefwechsel Wilhelm Ostwalds.* Berlin: Akademie-Verlag, 1961.

Lamirand, J. and Charles Brunold. *Cours de Chimie pour les Classes de Mathématiques-Spéciales.* Paris: Masson et Cie, 1927.

Langevin, Paul. 'L'oeuvre d'Henri Poincaré', *Revue de Métaphysique et Morale*, **21** (1913), 675–718.

'Sur la théorie du mouvement brownien', *Comptes rendus*, **146** (1908), 533–539.

de Lapparent, Albert Cocohon. *Traité de Geologie.* 2nd ed. Paris: F. Savy, 1885.

Laugier, Henry, Edmond Bauer, John Bernal, *et al. Hommage national à Paul Langevin et Jean Perrin.* Paris: Orléans, 1948.

Lebeau, Paul, and Pierre Girard. 'Hommage à Jean Perrin'. Paris: Hermann, 1950. 25 pp.

Le Chatelier, Henri. 'Les phénomènes de combustion et la production de la puissance mecanique de la chaleur et de l'électricité', *Revue Scientifique*, **9** (1898), 225–231.

Levy, Paul. 'Le Mouvement brownien', *Mémorial des sciences mathématiques*, Fasc. 126 (1954), 84 pp.

Livre d'hommage des lettres françaises à Émile Zola. Paris: Société libre d'Edition des Gens de Lettres, 1898.

Lot, Ferdinand. *Jean Perrin et les atomes.* Paris: Seghers (Savants du monde entier), 1963.

Mabilleau, Leopold. *Histoire de la philosophie atomistique.* Paris, 1895.

Mach, Ernst. *Erkenntnis und Irrtum.* Leipzig: J. A. Barth, 1905.

The Principles of Physical Optics. Trans. John Anderson and A. F. A. Young. New York: Dover, 1926.

The Science of Mechanics. Trans. Thomas J. McCormack. 6th ed. La Salle, Illinois: Open Court Publishing Co., 1960.

Maltezos, M. C. 'Sur le mouvement brownien', *Comptes rendus*, **121** (1895), 303–305.

Marx, Jarl Michael. 'Sur les mouvements des molécules', *Bibliothèque universelle des sciences, belles-lettres, et arts de Genève*, **48** (1831), 121.

Mecklenburg, Werner. *Die experimentelle Grundlegung der Atomistik.* Jena: Gustav Fischer, 1910.

Meyerson, Émile. *De L'Explication dans les sciences.* Paris: Payot, 1927.

Mieli, Aldo. 'Jean Perrin—*Les Atomes.* Theodor Svedberg—*Die Existenz der Moleküle.* M. Harten—*Die Hauptlehren der Averroes nach seiner Schrift.*

Geza Revesz—*Zur Grundlegung der Ton-psychologie.*' Genova: A. F. Formiggini, 1913.

Millikan, Robert A. *The Electron.* Chicago: University of Chicago Press, 1917.

Molisch, Hans. 'Über die Brownsche Molekularbewegung in Gasen, sichtbar gemacht durch ein gewöhnliches Mikroscop', *Zeitschrift für wissenschaftliche Mikroskopie*, **24** (1907), 97–103.

Muncke, Hofr. 'Ueber Robert Brown's mikrokopische Beobachtungen, über den Gefrierpunkt des absoluten Alkohols, und über eine sonderbare Erscheinung an der Coulomb'schen Drehwaage', Poggendorf's *Annalen der Physik*, **17** (1829), 159–165.

Nägeli, von. 'Ueber die Bewegungen kleinster Körperchen', *Sitzungsberichte der mathematisch-physikalischen Classe der k.b. Akademie der Wissenschaften zu München*, **9** (1879), 389-453.

New York Microscopical Society Journal, **5** (1889), 46. On fourteen-year-old Brownian movement.

Nernst, Walter. *Theoretische Chemie.* Stuttgart: Ferdinand Enke, 1921.

Ostwald, Wilhelm. 'Buchershau' [for The Svedberg's 'Studien zur Lehre von den Kolloiden Lösungen'], *Zeitschrift für physikalische Chemie*, **64** (1908), 508–512.

'Elements and Compounds', pp. 185–201 in C. S. Gibson and A. J. Greenaway (eds.), *Faraday Lectures, 1869–1928.* London: The Chemical Society, Burlington House, 1928.

Die Energie. 2nd ed. Leipzig: J. A. Barth, 1912.

'La Déroute de l'Atomisme contemporain', *Revue Générale des Sciences*, **6** (1895), 953–958.

'Lettre sur l'Énergétique', *Revue générale des sciences*, **6** (1895), 1068–1071.

'Vorbericht' to his *Grundriss der Physikalische Chemie.* Leipzig, 1908.

Painlevé, Paul. 'La philosophie de Marcelin Berthelot', *Revue du Mois*, **3** (1907), 513–522.

Paul, Harry. 'The Debate over the Bankruptcy of Science in 1895', *French Historical Studies*, **5**, #3 (1968), 299–327.

Perrin, Francis. 'Le mouvement brownien et la réalité moléculaire: Georges Gouy, Albert Einstein, Jean Perrin', *Institut de France. Académie des Sciences. Troisiéme Centenaire 1666–1966.* Paris: Gauthier-Villars, 1967, pp. 443–452.

'L'oeuvre scientifique de Jean Perrin', *Institut de France, Académie des Sciences (Publications Diverses)*, #14 (1962), pp. 6–18.

Perrin, Jean (*listed chronologically*)

*'Nouvelles propriétés des rayons cathodiques', *Comptes rendus*, **121** (1895), 1130.

*'Rayons cathodiques et rayons de Roentgen', *Annales de Chimie et de*

Physique, **11** (1897), 496–554. Thèse de Doctorat, Paris, juin 1897.

'Les hypothèses moléculaires', *Revue Scientifique*, **15** (1901), 449–461.

'Conditions qui déterminent le sens et la grandeur de l'électrisation par contact', *Comptes rendus*, **136** (1903), 1388–1391.

'Électrisation par contact (II)', *Comptes rendus*, **136** (1903), 1441–1443.

'Électrisation par contact (III)', *Comptes rendus*, **137** (1903), 513–514.

'Électrisation par contact (IV) et théorie des solutions colloïdales', *Comptes rendus*, **137** (1903), 564–566.

'Le principe d'équivalence et la notion d'énergie', *Revue de Métaphysique et de Morale*, (1903), 55–83.

Traité de Chimie Physique. Les Principes. Paris: Gauthier-Villars, 1903.

'Mécanisme de l'électrisation de contact et solutions colloïdales', *Journal de chimie physique*, **2** (1904), 601–651; **3** (1905), 50–110.

'Les rayons N existent-ils?', *Revue Scientifique*, **2** (1904), 622.

'Le contenu essentiel des principes de la thermodynamique', *Bullétin de la Société de Philosophie*, **6** (1906), 81–111.

'La discontinuité de la matière', *Revue du Mois*, **1** (1906), 323–344.

'À propos de "l'évolution des forces" ', *Revue du Mois*, **4** (1907), 606, 729; **5** (1908), 101.

'L'agitation moléculaire et le mouvement brownien', *Comptes rendus*, **146** (1908), 967–970.

'La loi de Stokes et le mouvement brownien', *Comptes rendus*, **147** (1908), 475–476.

'L'origine du mouvement brownien', *Comptes rendus*, **147** (1908), 530.

'Grandeur des molécules et charge de l'électron', *Comptes rendus*, **147** (1908) 594–596.

'Peut-on peser un atome avec précision', *Revue du Mois*, **6** (1908), 513–538.

'Mouvement brownien et molécules', *Journal de chimie physique*, **8** (1909), 57–91.

'Mouvement brownien et constantes moléculaires', *Comptes rendus*, **149** (1909), 477–479. Published with Dabrowski.

'Le mouvement brownien de rotation', *Comptes rendus*, **149** (1909), 459–551.

'Mouvement brownien et réalité moléculaire', *Annales de Chimie et de Physique*, **18** (1909), 1–114. Trans. Frederick Soddy, *Brownian Movement and Molecular Reality*. London: Taylor and Francis, 1910.

'Mouvement brownien et grandeurs moléculaires', *Le Radium*, **6** (1909), 353–362.

'Mouvement brownien et molécules', *Journal de physique théorique et appliquée*, **9** (1910), 5–39.

'Conférence sur le mouvement brownien', Paris: École professionelle supérieure des Postes et Télégraphes, 1910. 10 pp.

'Le mouvement brownien', Conférence à la Royal Institution. London, 24 fevrier 1911.

'La réalité des molécules', *Revue Scientifique*, **49** (1911), 774–784.

'Les déterminations des grandeurs moléculaires', *Comptes rendus*, **152** (1911), 1165–1168.

*'Les grandeurs moléculaires (nouvelles mesures)', *Comptes rendus*, **152** (1911), 1380–1382.

*'L'agitation moléculaire dans les fluides visqueux', *Comptes rendus*, **152** (1911), 1569–1571. Published with Niels Bjerrum.

'Les preuves de la réalité moléculaire', pp. 1–53 in *Les idées modernes sur la constitution de la matière*. Paris: Gauthier-Villars, 1913.

'Les preuves de la réalité moléculaire', pp. 153–251 in *La théorie du rayonnement et les quanta* (eds.) Paul Langevin and Maurice de Broglie. Paris: Gauthier-Villars, 1912.

'L'agitation moléculaire', *Revue du Mois*, **14** (1912), 513–540.

Les Atomes. Paris: Alcan, 1913. 4th ed., 1914. *Atoms*. Trans. D. L. Hammick. 2nd rev. English ed. New York: Van Nostrand, 1923.

'Observations sur les lames minces', Conférence à la seance annuelle de la Société suisse de Physique, and at the Société de Chimie physique à Paris, avril 1913.

'Compressibilité osmotique des émulsions considerées comme fluides à molécules visibles', *Comptes rendus*, **158** (1914), 1168.

'La stratification des lames liquides', *Annales de Physique*, **10** (1918), 160–184.

Notice sur les Titres et Travaux Scientifiques. Paris: Gauthier-Villars, 1918.

'La fluorescence', *Annales de Physique*, **10** (1918), 133–159.

'Matière et lumière', *Annales de Physique*, **11** (1919), 1–108.

*'L'origine de la chaleur solaire', *Scientia* (1921), 345.

Notice sur les Travaux Scientifiques. Toulouse: Edouard Privat, 1923.

'La chimie physique', pp. 18–28 in Perrin, Langevin, Urbain, *et al.*, *L'orientation actuelle des sciences*. Paris: Bibliothèque de philosophie contemporain, 1930.

Les Éléments de la Physique. Paris: Albin Michel, 1930.

'La structure discontinue de la matière', Conférence Nobel, Stockholm, 12 decembre 1926. Stockholm: Imprimerie Royale. P.A. Norstedt & Fils, 1927. 29 pp.

'La Recherche Scientifique'. *Actualités scientifiques et industrielles*, #58; Paris: Hermann, 1933. 25 pp.

*'L'Organisation de la Recherche Scientifique en France', Paris: Hermann, 1938. 50 pp.

La Science et l'Espérance. Paris: Presses Universitaires de France, 1948. A number of speeches collected in this volume have been used as sources, including 'H. G. Wells', *Bullétin de l'Association France–Grande-Bretagne*, #65 (1927), 16; 'L'astronomie libératice de la raison humaine', preface to Paul Couderc, *L'Architecture de l'Univers*. Paris: Gauthier-Villars, 1930; 'La science libérera l'homme', *Le Petit Journal*, 19 mars 1937; and *'Le Congrès du Palais de la Découverte', *Bullétin des Congrès*, #26, 6–13.

Oeuvres scientifiques de Jean Perrin. Paris: Centre National de la Recherche Scientifique, 1950.

Picard, Émile. *La science moderne et son état actuel*. Paris: Flammarion, 1905.

La vie et l'oeuvre de Pierre Duhem. Paris: Gauthier-Villars, 1922.

Planck, Max. *The New Science*. Trans. James Murphy. New York: Meridian, 1959.

Vorlesungen über Thermodynamik. Leipzig: Verlag von Veit & Co., 1897.

Poincaré, Henri. 'Éloge to Ludwig Boltzmann', *Comptes rendus*, **143** (1906), 997.

'Les rapports de la matière et de l'éther'. Extract from *Journal de Physique*, mai 1912. Paris: Deslis Frères, 1912. 16 pp.

Science and Hypothesis. Trans. W. J. G. New York: Dover, 1952.

'Sur la théorie cinétique des gaz', *Revue générale des sciences*, **5** (1894), 513–521.

Thermodynamique. Cours de physique mathématique. 2nd ed. Paris: Gauthier-Villars, 1908.

'The Principles of Mathematical Physics', trans. George B. Halsted, pp. 604–622 in *Congress of Arts and Science. Universal Exposition. St. Louis, 1904*, (ed.) Howard J. Rogers. New York: Houghton, Mifflin & Co., 1905.

The Value of Science. Trans. George B. Halsted. New York: Science Press, 1907.

Poincaré, Lucien. 'L'évolution de la physique', *Revue scientifique*, **5** (1906), 481–486.

Puccianti, L. and E. Vigezzi. 'Sulla Osservazione della particelle ultramicroscopiche', *Archivio di Fisologia*, **2** (1904), 307–320.

Quincke, G. 'Ueber die Bewegung und Anordnung kleiner Teilchen, welche in Flüssigkeiten schweben', *Verhandlungen der Gesellschaft deutsche Naturforscher und Ärtze*, **2** (I), (1898), 26–29.

Ranc, Albert. *Jean Perrin, un grand savant au service du socialisme*. Pref., Léon Blum. Paris: Editions de la Liberté, 1945.

Raspail. 'Note on Mr. Brown's Microscopical Observations on the Active Molecules of Organic and Inorganic Bodies', *Edinburgh New Philosophical Journal*, **8** (1830), 43–44.

'Observations and Experiments tending to demonstrate that the granules which are discharged in the explosion of a grain of pollen, instead of being analogous to spermatic animalcules, are not even organised bodies', *Edinburgh Journal of Science*, **10** (1829), 96–106.

Review of Jean Perrin's *Les Atomes*, *Nature*, **91** (March-August 1913), 473.

Review of Ernst Mach's *Science of Mechanics*, *Revue Scientifique*, **20** (1903), 787–788.

Rey, Abel. *L'énergétique et le mécanisme*. Paris: Alcan, 1907.

La philosophie moderne. Paris: Flammarion, 1908.

'Les principes philosophiques de la chimie physique', *Revue philosophique de la France et de l'Étranger*, **57** (1904), 393–409.

Robin, Charles Philippe. *Traité du Microscope*. Paris, 1871.

Sageret, Jules. *La Vague Mystique*. Paris: Flammarion, 1920.

Sarton, George. 'Paul, Jules and Marie Tannery', *Isis*, **38** (1947), 33–51.

Seddig, Max. 'Über die Messung der Temperaturabhängigkeit der Brownschen Molekularbewegung', *Physikalische Zeitschrift*, **9**, #14 (1908), 465–468.

Siedentopf, H. F. 'On the rendering visible of ultramicroscopic particles and of ultramicroscopic bacteria', *Royal Microscopical Society Journal*, 17 June 1903, 573–578.

and R. A. Zsigmondy. 'Sichtbarmachung und Grossenbestimmung ultramikroskopischer Teilchen, mit besonderer Andwendung auf Goldrubinglaser', *Annalen der Physik*, **10** (1903), 1–39.

Svedberg, Theodor. *Die Existenz der Moleküle*. Leipzig: Akademische Verlagsgesellschaft m.b.H., 1912.

Studien zur Lehre von den kolloiden Lösungen. Uppsala: Nova Acta Regiae Societatis Scientiarum Upsaliensis (4), **2**, #1, 1907.

'Über die Eigenbewegung der Teilchen in kolloidalen Lösungen', *Zeitschrift für Electrochemie*, **12** (1906), 853–860, 909–910.

Taton, René. *Causalités et Accidents de la Découverte Scientifique*. Paris: Masson et Cie, 1955.

Thirion, J. 'Les Mouvements Moléculaires', *Revue des questions scientifiques*, **7** (1880), 5–55.

Thorpe, T. E. *Essays in Historical Chemistry*. London: Macmillan, 1894.

Thulin, Elsa. 'Souvenirs sur Jean Perrin', Stockholm: Bonnier for Edition de l'Institut français, 1949. 16 pp.

Tyndall, John. 'Atoms and Equivalents', *Nature*, **16** (1877), 293–294.

Urbain, Georges. 'Le rôle de l'atomistique dans l'enseignement et dans la recherche', *Revue du Mois*, **7** (1909), 417–429.

van Melsen, Andrew G. *From Atmos to Atom*. New York: Harper Torchbooks, 1960.

Virtanen, Reino. 'Marcelin Berthelot. A Study of a Scientist's Public Role', *University of Nebraska Studies*, new series #31 (April 1965), 67 pp.

Washburn, Edward W. *Introduction aux principes de la chimie physique du point de vue de l'atomistique et de la thermodynamique moderne.* Trans. Henri Weiss et Albert Noyes. Paris: Payot, 1925.

Weinberg, Carlton Berenda. *Mach's Empirio-Pragmatism in Physical Science.* New York: Albec Press, 1937.

Whewell, William. *The Philosophy of the Inductive Sciences.* 2 vols. London: John W. Parker, 1847.

Wiener, Christian. 'Erklärung des atomischen Wesens des tropbarflüssigen Körperzustandes, und Bestätigung desselben durch die sogenannten Molekularbewegungen', *Annalen der Physik,* **118** (1863), 79–94.

Wurtz, Charles Adolphe. *La Théorie Atomique.* Paris: Librairie Germer Baillière et Cie, 1879.

Zsigmondy, R. A. *Kolloidchemie* (1912), trans. Ellwood B. Spear, pp. 1–28 in *The Chemistry of Colloids.* New York: John Wiley and Sons, 1917.

Other sources on the life and work of Jean Perrin which might be consulted, but which have not been directly cited in this analysis, are:

Heathcote, Niels, 'Jean Perrin', pp. 249–258 in *Nobel Prize Winners in Physics, 1901–1950.* New York, 1953.

Hylleraas, Egil. 'Jean Perrin', *Fra fysikkens verden,* Arg.1 2, hefte 1 (1950), 6 pp.

Townsend, J. S. 'Jean Baptiste Perrin', *F.R.S. Obituary Notices,* **4** (1942–1944), 301–305.

There are also short articles in the *Journal de chimie physique,* **39** (1942), p. 45; *Renaissance,* **1** (1942); and *Technique moderne,* **29** (1937), p. 11.

Unpublished Sources

The following entries, with a few exceptions, refer to letters which were written to Jean Perrin and which were most generously loaned to me by his son. The exceptions are starred and cite correspondence obtained in photocopy for me by Professor Hiebert from the collection of Einstein correspondence at the Princeton Institute for Advanced Study. In addition to the correspondence which Professor Perrin allowed me to use, I have drawn upon conversation with him (29 October 1968), as well as his then unpublished paper, 'Jean Perrin, prix Nobel de Physique, 1926', 19 pp.

Branly, E. Paris, 30 January 1909.

Crookes, William. 7 Kensington Park Gardens, London W., 24 October 1910.

Einstein, Albert. Zurich, 11 November 1909.
  Zurich, 12 January 1911.
  *5 November 1919.

Gouy, Léon. Lyon, 31 October 1908.
  Lyon, 18 March (probably 1910).

Guye, Philippe A. 1 February 1913.
*Hopf, Ludwig [to Albert Einstein]. Tuesday (probably from Zurich, December 1910).
Kammerlingh-Onnes, H. Leiden, 19 January 1910.
Keesom, W. H. Leiden, 29 November 1911.
Kurlbaum, F. Berlin, 25 April 1911.
[Letter with illegible signature.] Utrecht, 23 September 1909.
Loeb, Jacques. Rockefeller Institute for Medical Research. Marine Laboratory, Woods Hole, Mass., 7 August 1918.
Lorentz, H. A. Leiden, 14 January 1910.
Mecklenburg, Werner. Clausthal im Harz, 29 January 1910.
Olivier, Louis. 29 September 1908.
Ostwald, Wilhelm. 25 April 1913.
Ostwald, Wolfgang. Leipzig, Brandworwekst 77, 14 October 1909.
*Perrin, Jean [to Albert Einstein]. Paris, 28 August 1919.
Picard, Émile. 30 November 1909.
Poincaré, Lucien. 4 February 1913.
Poynting, J. H. 20 February 1910.
Ramsay, William. 19 Chester Terrace, Regent's Park, N.W., 4 March 1912.
Rubens, H. Berlin, 12 October 1910.
Institut, 24 October 1910.
Rutherford, Ernest. 17 Wilmslow Road, Withington, Manchester, 4 May 1909.
Smoluchowski, Maryan. Lernberg (Antrich) Dlugoss 8, 24 January 1909.
Soddy, Frederick. The University, Glasgow, 11 January 1909.
Thomson, J. J. Cambridge, 10 February 1913.

Index

BIBLIOGRAPHY